WILD DOGS

The Natural History of the
Nondomestic Canidae

The Natural History of the Nondomestic Canidae

by
JENNIFER W. SHELDON

With an introduction by
Patricia D. Moehlman, Wildlife Conservation International

ACADEMIC PRESS, INC.
HARCOURT BRACE JOVANOVICH, PUBLISHERS
San Diego   New York   Boston   London   Sydney   Tokyo   Toronto

Copyright © 1992 by ACADEMIC PRESS, INC.

All Rights Reserved.

No part of this publication may be reproduced or transmitted in any form or by any means, electronic or mechanical, including photocopy, recording, or any information storage and retrieval system, without permission in writing from the publisher.

Academic Press, Inc.
San Diego, California 92101

United Kingdom Edition published by
Academic Press Limited
24–28 Oval Road, London NW1 7DX

Library of Congress Cataloging-in-Publication Data

Sheldon, Jennifer W.
    Wild dogs : the natural history of the nondomestic Canidae /
Jennifer W. Sheldon.
        p.   cm.
    Includes bibliographical references and index.
    ISBN 0-12-639375-3
    1. Canidae.   I. Title
QL737.C22S535   1991
599.74'442--dc20   91-14127
                                                                                       CIP

PRINTED IN THE UNITED STATES OF AMERICA
91 92 93 94   9 8 7 6 5 4 3 2 1

This book is dedicated to Ian Gersten

TABLE OF CONTENTS

PREFACE

Of all the animals, canids are the group with which humans have had the most longstanding, universal, and profound associations. These associations, both affectionate and adversarial, reach back into prehistoric time. For reasons about which we can only speculate, dogs were the first animals to be domesticated, thus our continued fascination with wild cousins of domestic dogs is deep-rooted. Wild canids are important for a number of reasons beyond their historical ties to humans. Representatives occur throughout the world from arctic regions to tropical forests. A number of species are economically important as furbearers, and others compete with humans for access to large ungulates or domestic stock. From ecological and behavioral standpoints, canids are distinguished by their diversity. Social organization among the Canidae ranges from solitary to among the most highly social of all mammals. Canid adaptability and behavioral plasticity are remarkable, both on a group and an individual level. The status of over half the wild dog species is endangered, threatened, or unknown. Attention to their ecology is now of critical importance: decisions taken in the next few years will seal the fate of many of these species.

For these reasons, the information in this volume is of critical importance. Until the publication of this book, there has been no single source which presents a comprehensive, current natural history of the nondomestic dog species. In this book, a prodigious amount of previously uncollected information is presented in a straightforward form. The book is intended as a general reference work. Biologists, wildlife managers, mammalogists, conservationists, students, and carnivore specialists will find here information assembled nowhere else. Over 600 sources are included in the bibliography, so the book also serves as an entry to the literature for those seeking more technical or specialized knowledge. Naturalists and outdoorsmen will also enjoy discovering the particulars of familiar and unfamiliar canid species. Although carnivores are typically difficult to observe in the wild because of rarity and reticence, a number of wild dog species are readily studied by amateur naturalists in sign, scat, by sight, and through their prominent long-range vocalizations.

The organization of the book is alphabetical by genus, and, within each genus, alphabetically by Linnean species name. Those unfamiliar with the Latin names will find common names listed in the Table of Contents, in the table on p. 6, and at each section heading. Most canid species suffer from a profusion of common names, thus the use of Latin names is the clearest designator: I hope the nonspecialist will not be put off by their prominence. Within each section of the book there are seven subheadings, each summarizing the current state of knowledge on the following seven topics: distribution and habitat, physical characteristics, taxonomy, diet, activity, reproduction, and social organization and behavior. Thus the book can be consulted across species in order to compare information on particular aspects of natural history. In some cases, very little is known about a species. In other cases, the amount of available information is enormous, and has been distilled to summary form. The volume is intended as a straightforward assemblage of material. It points the way toward, but is not intended to provide, a synthetic or theoretical big picture. As the title indicates, this volume deals only with nondomestic canid species. Obviously, domestic dogs and their feralized variants require a volume of their own.

Because of the ever accelerating pace of biological destruction worldwide, any wildlife study is concerned with conservation de facto. This is nowhere more true than in the study of large carnivores, who are typically the first to be overwhelmed by human persecution and habitat destruction. As I wrote this book, it became appallingly apparent that human actions have pushed many wild dog species to the brink of extinction. It is my hope that all future canid studies operate with this knowledge as a guiding principle.

The kind assistance and encouragement of the following people has been invaluable to me. I thank each one of them: Marc Bekoff, Mark Boyce, Jerry Choate, Kim Fadiman, Jeff Foott, Jeff Holtmeier, Ted Kerasote, Peter Klopfer, Phil Lehner, David Macdonald, Patricia Moehlman, Skip Sheldon, and Suzanne and Stewart Wolff.

Jennifer W. Sheldon
P.O. Box 1035
Jackson, WY 83001

Introduction
by Patricia D. Moehlman*

Family Canidae consists of approximately 35 species that are categorized into 15 genera. Typically they are lithe runners and opportunistic omnivores. They range in body weight from 1.5 to 31.1 kg and show great diversity in diet, habitat, and distribution (Gittleman, 1984). Family Canidae has a suite of characteristics that are unique or unusual within Carnivora and Mammalia. These characteristics include the following: long-term monogamy (Kleiman, 1977); paternal investment; comparatively large litters and a long period of infant dependency (Kleiman and Eisenberg, 1973); and a cooperative breeding system in which some offspring remain in the natal family and provision and protect their younger relatives (Moehlman, 1986). Family members share food and provide care for sick or injured relatives.

Canidae have interspecific behavioral trends that correlate with body weight (Moehlman, 1986). For example, small canids (those less than 6 kg) such as red foxes (*Vulpes vulpes*) and bat-eared foxes (*Otocyon megalotis*), are solitary foragers, are usually monogamous but have a tendency toward polygyny, have an adult sex ratio biased toward females, male dispersal, and some non-reproductive females helping to rear pups. Medium-sized canids (6–13 kg) including silverbacked [black-backed] and golden jackals (*Canis mesomelas* and *C. aureus*) and coyotes (*C. latrans*) are facultative cooperative foragers, tend to be more strictly monogamous, have equal adult sex ratios, and both males and females help to raise younger relatives and/or disperse. Most large canids (greater than 13 kg) such as African hunting dogs (*Lycaon pictus*) are obligatory cooperative hunters, tend to have a monogamous mating system with indications of polyandry, have adult sex ratios skewed toward males, have male helpers, and have female emigration. These are general trends within Canidae, but there are in-

*Wildlife Conservation International, New York Zoological Society, Bronx Zoo, Bronx, NY 10460.

1

teresting and important exceptions interspecifically, and there is a great deal of variation within species.

As body weight increases in canids there are also significant and intriguing correlations with important life history traits (Bekoff *et al.*, 1981; Gittleman, 1984, 1985; Moehlman, 1986). Similar to other mammals, canid neonate weight increases with maternal body weight at approximately a 3/4 exponent of female weight. However, canids appear to be unique among mammals in showing a trend toward larger litter sizes and weights as maternal body weight increases. Thus as canids get larger, not only do the relatively smaller and less-developed pups potentially require greater parental investment after birth, but in addition there are more neonates to care for. Canidae tend to have the heaviest litter weights and the shortest gestation lengths within Carnivora (Gittleman, 1984, 1985). Thus female canids must contend with large prepartum investment, and as body weight increases require more assistance in feeding and defending their larger litters of proportionately smaller offspring.

Such physiological constraints are consistent with the general behavioral trends observed in Canidae and they provide insight into parental investment and sexual selection in canid social systems. For example, smaller females will have fewer and relatively heavier pups that potentially require less paternal investment after birth. The potential arises for a male to invest in the offspring of more than one female, and polygyny is possible. Parental investment and sexual selection theory would then predict that as fathers contribute less, there would be reduced competition by females for males, there would be a tendency toward polygyny, and males would disperse (Trivers, 1972). This set of behaviors has been observed in small canids like kit foxes (*Vulpes macrotis*), arctic foxes (*Alopex lagopus*), bat-eared foxes (*Otocyon megalotis*), and red foxes (*Vulpes vulpes*) (Egoscue, 1962; Hersteinsson, 1984; Nel, 1978; Nel *et al.*, 1984b; Lamprecht, 1979; Macdonald, 1979a, 1980, 1981; Brady, 1978, 1979; Moehlman, 1986). The availability of food and the ability of the male to control resources could affect the relative costs and benefits of maternal and paternal investment and could determine whether the mating system is monogamous or polygynous.

By contrast, large canids usually produce larger litters of relatively less-developed pups. Females make a large investment in their offspring before birth, and require substantial male involvement in their pups after birth. If their pups are to survive, females cannot afford to share this investment with other females (e.g., polygyny is unlikely) and competition for males can be intense. One would therefore expect a significant bias in the pup and adult ratios toward males, fierce competition between females for dominance, female emigration, and male relatives helping. This is the pattern observed in African hunting dogs (*Lycaon pictus*) (Frame and Frame, 1977; Malcolm, 1979; Frame *et al.*, 1979; Malcolm and Marten, 1982).

Such correlations of mating and breeding strategies with possible physiological constraints of body size and numbers of dependent offspring can be confounded by ecological factors. Ecological constraints, particularly the size and the spatial and temporal availability of food, can affect territory size, group size, and reproductive strategies. This effect can be seen in several canid species that have litter sizes that are inconsistent with other species of similar weight. Arctic foxes, for example, can have unusually large litters for their maternal weight category, and ecological factors appear to have a major impact on their reproductive effort. In northwestern Canada, where lemmings (*Dicrostonyx torquatus* and *Lemmus sibiricus*) can be very abundant during the whelping season, the mean litter size of arctic foxes is 10.1 (MacPherson, 1969). By contrast, in coastal areas of Iceland, where food availability is patchy but relatively steady in abundance, the mean litter size is 4.0 (Hersteinsson, 1984). Arctic foxes seem to be capable of exploiting periodically abundant food resources and dramatically increasing their reproductive rate.

Another unusual species is the maned wolf (*Chrysocyon brachyurus*), a large canid (about 23 kg) whose mean litter size of two pups is the lowest in the Canidae. It is the only large canid that forages primarily on rodents and fruit (Dietz, 1984). Feeding on items of food that are relatively small may impose energetic constraints on the female.

Research on the varying ecological and demographic circumstances that a single species contends with offers the opportunity to investigate linkages between physiology, ecology, and behavior. Field studies have revealed that canid species are capable of an impressive degree of intraspecific variation both between and within populations as the ecological parameters vary through time. Further research is needed to understand how ecological factors and physiological constraints interact to determine canid reproductive strategies and social organization.

A Note on Taxonomy

Current views on the taxonomy of the Canidae are diverse, with a number of major ongoing disputes. The division into subfamilies is still unsettled; therefore, usage has not stabilized (Anderson and Jones, 1984). The taxonomic arrangement in this work is moderately conservative and, for the most part, agrees with that of Nowak and Paradiso (1983) and Berta (1987). For alternate taxonomies see Ginsberg and Macdonald (1990), Van Gelder (1978), Anderson and Jones (1984), Clutton-Brock *et al.* (1976), Macdonald (1984), Langguth (1975b), Osgood (1934), Cabrera (1931), or Stains (1975).

The following species are not recognized in this text:

Dusicyon fulvipes: Darwin's or chiloe fox. Found on an island off the west coast of Chile and immortalized by Darwin's mention in the "Voyage of the Beagle" (1962) where he gives an account of sneaking up behind one and knocking it on the head with a geological hammer. Probably conspecific with *Pseudalopex griseus*. (See *P. griseus* for details.)

Dusicyon inca: The Peruvian fox. Found in southern Peru. Described from a single skull and skin. Perhaps conspecific with *Pseudalopex culpaeus* or *P. griseus*.

Dusicyon culpaeolus: Santa Elena fox. Occurs in southeastern Uruguay. Very similar to *P. culpaeus*, though smaller. Bears a close physical resemblance to *P. gymnocercus* (Clutton-Brock *et al.*, 1976).

Dingoes (*Canis familiaris dingo*) were introduced to Australia in recent prehistoric times. Similarly, New Guinea wild dogs (*Canis familiaris hallstromi*) are probably descended from domestic dogs and are closely related to dingos (Anderson and Jones, 1984; Simonsen, 1976; Nowak and Paradiso, 1983). These species are not included here because this work is only concerned with nondomestic species of the Canidae.

In the past, domestic dogs were usually referred to as *Canis familiaris*. More recently, the designation of *Canis lupus familiaris* has been used. Since wolves are now thought to be the recent ancestors of all domestic dog varieties (Simonsen, 1976; Clark *et al.*, 1975; Wayne and O'Brien, 1987), this designation has a great deal of merit. But this question of nomenclature has not been resolved, and usage still varies.

Family Canidae Species List: 35 Species in 15 Genera[a]

Genus and Species	Taxonomist	Common Name
Alopex	**Kaup, 1829**	
A. lagopus	(Linnaeus, 1758)	Arctic fox
Atelocynus	**Cabrera, 1940**	
A. microtis	(Sclater, 1882)	Small-eared dog
Canis	**Linnaeus, 1758**	
C. adustus	Sundevall, 1846	Side-striped jackal
C. aureus	Linnaeus, 1758	Golden jackal
C. latrans	Say, 1823	Coyote
C. lupus	Linnaeus, 1758	Gray wolf
C. mesomelas	Schreber, 1778	Black-backed jackal
C. rufus	Audubon and Bachman, 1851	Red wolf
C. simensis	Rüppell, 1835	Ethiopian jackal
Cerdocyon	**Hamilton Smith, 1839**	
C. thous	(Linnaeus, 1766)	Crab-eating fox
Chrysocyon	**Hamilton Smith, 1839**	
C. brachyurus	(Illiger, 1815)	Maned wolf
Cuon	**Hodgson, 1838**	
C. alpinus	Pallas, 1811	Dhole
Dusicyon	**Hamilton Smith, 1839**	
D. australis	(Kerr, 1792)	Falkland Island wolf
Fennecus	**Desmarest, 1804**	
F. zerda	Zimmermann, 1780	Fennec fox
Lycaon	**Brookes, 1827**	
L. pictus	(Temminck, 1820)	African wild dog
Nyctereutes	**Temminck, 1839**	
N. procyonoides	(Gray, 1834)	Raccoon dog
Otocyon	**Muller, 1836**	
O. megalotis	Desmarest, 1822	Bat-eared fox
Pseudalopex	**Burmeister, 1856**	
P. culpaeus	(Molina, 1782)	Culpeo
P. griseus	(Gray, 1834)	Chilla
P. gymnocercus	(Fischer, 1814)	Pampas fox
P. sechurae	(Thomas, 1900)	Sechura fox
P. vetulus	(Lund, 1842)	Hoary fox
Speothos	**Lund, 1839**	
S. venaticus	Lund, 1842	Bush dog

Genus and Species	Taxonomist	Common Name
Urocyon	**Baird, 1858**	
U. cinereoargenteus..	(Schreber, 1775)	Gray fox
U. littoralis	(Baird, 1858)	Island gray fox
Vulpes	**Bowdich, 1821**	
V. bengalensis	Shaw, 1800	Bengal fox
V. cana	Blanford, 1877	Blanford's fox
V. chama	A. Smith, 1834	Cape fox
V. corsac	Linnaeus, 1768.............	Corsac fox
V. ferrilata	Hodgson, 1842	Tibetan sand fox
V. macrotis	Merriam, 1888	Kit fox
V. pallida	(Cretzschmar, 1826)	Pale fox
V. rüppelli	Schinz, 1825	Rüppell's fox
V. velox	(Say, 1823)	Swift fox
V. vulpes	Linnaeus, 1758.............	Red fox

[a]For Linnean (Latin) synonymy with noncanid mammalian names see Nowak and Paradiso (1983).

Arctic Fox *(Alopex lagopus)*
Credit: Thomas Mangelsen/Images of Nature

CHAPTER 1
Genus *Alopex*

Alopex lagopus: Arctic Fox

Small, mobile, and omnivorous, arctic foxes are one of the few species of mammals that have successfully adapted to the forbidding climate of the arctic regions. They occur over a vast circumpolar range and have developed a variety of regional adaptations, as well as exhibiting an overall versatility and adaptability. Individuals make the most extensive movements of any terrestrial mammal except for man (Nowak and Paradiso, 1983). They are commercially very important as furbearers, and are trapped, snared, shot, and raised in captivity. Economically they are one of the more important resources of the Canadian arctic regions (Banfield, 1974). During the 1976–77 season 36,500 pelts were marketed in Canada and 4,200 in Alaska (Deems and Pursley, 1978, cited in Nowak and Paradiso, 1983). The annual harvest worldwide is on the order of 100,000 foxes (Garrott and Eberhardt, 1987). Despite their commercial importance and their innate lack of wariness of humans, little is known about the social organization of free-ranging arctic foxes. They are neither solitary nor highly social.

DISTRIBUTION AND HABITAT

Distribution is holarctic circumpolar, including the arctic or tundra regions of North America, Eurasia, Scandinavia, Spitzbergen, Greenland, Iceland, and islands of the Arctic, North Atlantic, and North Pacific Oceans (Banfield, 1974; Chesemore, 1975; Ewer, 1973). There is a relict population in the Altai Mountains in Central Asia (Banfield, 1974).

Arctic and alpine tundra and coastline habitats are favored. Arctic foxes are highly mobile and are good swimmers (Novikov, 1962). Marked individuals have been found over 1,500 km from their original point of capture (Wrigley and Hatch, 1976, cited in Nowak and Paradiso, 1983). Regular winter migrations take place and large-scale emigrations may follow drastic reductions in food supplies. Fall and winter movements of arctic foxes to Prudhoe Bay from a distance of up to 1,000 km have been reported (Eberhardt *et al.*, 1983b). In some regions foxes move inland to the forests, in others, far out on the sea ice. Tracks have been seen as far north as 88 degrees north latitude, and foxes have been sighted more than 450 km from the nearest ice-free land off Greenland (Banfield, 1974; Novikov, 1962; Chesemore, 1975). Mass emigrations have been observed in some areas as well. Food shortages may trigger these unidirectional movements (Chesemore, 1975). The timberline usually forms the natural southern boundary of their range, although they may penetrate deeply into boreal forest zones during winter (Banfield, 1974; Stroganov, 1962).

Dens are generally found on the open tundra. Arctic foxes may also make dens among dunes or in pingos, among rock in talus slopes, in rock fields or similar rocky areas; they also use the abandoned burrows of Siberian marmots. On the open tundra, dens have a variety of forms but are usually a mound 1–4 m high. Some dens are small with few entrances, while others have an extensive network of entrances and tunnels covering anywhere between 30 and 180 m². The number of den entrances may range from 4 to 100 in very old dens. Old, large dens have been used for many generations and in some cases for centuries (MacPherson, 1969). Some arctic foxes return to the same den in successive years. Dens often "radically alter the floristic community near the burrow" (Chesemore, 1975, p. 157): Den entrances are characterized by abundant vegetation due to aeration of the soil and addition of organic material. Dens may be used throughout the year in some regions and only during denning season in others. In winter or during blizzards, foxes shelter in burrows dug in the snow (Eberhardt, 1976; Chesemore, 1975; Banfield, 1974; Novikov, 1962).

Arctic foxes are attracted to areas of human activity, such as camps and construction sites, where they find food and shelter. Except in areas where they are intensely hunted by man, they are very tolerant of human activity. In fact, they are not wary of humans, and may steal things (Eberhardt, 1976; Banfield, 1974; Pedersen, 1975).

In areas where their distribution overlaps, arctic foxes actively avoid red foxes (*Vulpes vulpes*) (Schamel and Tracy, 1986). In captivity, red foxes dominate arctic foxes (Rudzinski *et al.*, 1982). There is competition for food and den sites, as well as outright predation by red foxes on *Alopex*. Red foxes may represent the greatest single threat to arctic fox populations in Scandinavia (Hersteinsson *et al.*, 1989). The eradication of wolves (*Canis lupus*) throughout much of their former range may have an adverse impact on arctic foxes. Wolf kills, which once provided a high quality source of scavenged food for arctic foxes, are no longer available. Wolves also may have effectively depressed red fox populations. Without wolves, red fox populations may have rebounded, their increased numbers acting to depress arctic fox populations (Hersteinsson *et al.*, 1989). In the low arctic regions where distribution of wolves, red foxes, and arctic foxes overlaps, the complexity of their interactions is evident, although as yet not clearly understood.

PHYSICAL CHARACTERISTICS

Short extremities and tremendously thick fur are among the adaptations to the demanding climate. Weights vary greatly, ranging from 1.4 kg to a maximum of 9 kg (Chesemore, 1975; Pedersen, 1975; Novikov, 1962). There is slight sexual dimorphism, with males being larger and heavier than females. The average weight of Canadian adult males is 3.5 kg (range 3.2–4.0 kg) and that of females is 2.9 kg (range 2.5–3.3 kg) (Banfield, 1974). Hersteinsson and Macdonald (1982) give slightly higher average weights—3.8 kg for males and 3.09 kg for females. Head-plus-body length of males averages 55 cm; that of females, 53 cm (Hersteinsson and Macdonald, 1982). Tails are long, over half the head-plus-body length, ranging from 26 to 34 cm (Garrott and Eberhardt, 1987; Banfield, 1974; Hersteinsson and Macdonald, 1982; Novikov, 1962).

The fur, particularly the winter coat, is soft and thick with dense underfur and long, fine guard hairs. It has the best insulation value of any mammal fur, including that of polar bears, wolves, and grizzly bears (Ewer, 1973; Hersteinsson and Macdonald, 1982; Chesemore, 1975; Novikov, 1962; Clutton-Brock *et al.*, 1976). There are two distinct color morphs. This dichromatism seems to be controlled at a single genetic locus, the white being recessive, the blue, dominant (Slagsvold, 1949, cited in Hersteinsson and Macdonald, 1982; Banfield, 1974). The ratio of blue to white forms in different populations varies greatly (Banfield, 1974; Hersteinsson and Macdonald, 1982; Chesemore, 1975). Pelage color of both blue and white arctic foxes differs seasonally, and for both morphs, there are two molts a year, in spring and autumn (Banfield, 1974; Chesemore, 1975; Ewer, 1973; Stroganov, 1962;

Hersteinsson and Macdonald, 1982). The white form molts to become pure white in winter; in summer it is anywhere from dark brown, brown, or brownish-gray to smoky gray, with lighter underparts. The basic color of the blue-phase individuals is even more varied. Pelage ranges from pearl gray, light blue–gray, gray, gray–brown, or dark brown to almost black. Seasonal color changes are considerably less dramatic than in the white form; the winter coat is slightly lighter than the summer one (Banfield, 1974; Chesemore, 1975; Hersteinsson and Macdonald, 1982; Stroganov, 1962; Novikov, 1962; Clutton-Brock et al., 1976).

The muzzle is short and blunt, the rostrum is relatively broad, and the ears are short and do not protrude much above the winter coat. The pupils are elongate with golden yellow irises (Hall and Kelson, 1959; Chesemore, 1975; Stroganov, 1962; Novikov, 1962; Clutton-Brock et al., 1976). The soles of the feet are thickly furred, and the toe pads are completely covered by thick hair—an attribute that gave rise to the specific Linnaean name *lagopus* which means "hare-footed."

Skull dimensions overall are shorter and flatter than those of red foxes. The braincase exceeds the facial region in length, which gives a short-muzzled aspect to the head. The facial region is short and broad. The canines are relatively short and weak, as is the bite, and the teeth of the lower jaw are closely spaced (Clutton-Brock et al., 1976; Novikov, 1962; Stains, 1975; Stroganov, 1962). The general form of the skull is intermediate between that of *Canis* and *Vulpes* (Clutton-Brock et al., 1976). The dental formula conforms to the usual canid pattern: incisors 3/3, canines 1/1, premolars 4/4, molars 2/3 = 42. See Stains (1975) for remarks on skull differences between *Vulpes* and *Alopex*.

TAXONOMY

In the past some taxonomists have divided *Alopex* into a number of different species, but more recently a single species with circumpolar distribution is recognized. Van Gelder (1978) recognized *Alopex* as a subgenus of *Canis*. Bobrinskii (1965, cited in Nowak and Paradiso, 1983) regarded *Alopex* as a subgenus of *Vulpes*. Others have placed the arctic fox within the genus *Vulpes* (Youngman, 1975, cited in Nowak and Paradiso, 1983). Recognizing that arctic foxes resemble *Vulpes* closely in physical characteristics, Clutton-Brock et al. (1976) none-theless recommend separate monospecific generic status. Nowak and

Paradiso (1983) also grant full generic status to *Alopex*. The karyotype of arctic foxes is highly distinctive 2n = 48–50: NF = 94 (for details see Wurster and Benirschke, 1968). There are four subspecies (Garrott and Eberhardt, 1987).

DIET

Arctic foxes are omnivorous, opportunistic feeders. Diet composition varies from region to region, season to season, and year to year. In most regions small rodents, primarily lemmings, are most important year-round. Other small mammals, such as voles, ground squirrels, and young hares, are also featured in the diet. Birds, eggs, nestlings, and fledglings are all major food sources, and bird breeding colonies are favored feeding areas. Ptarmigan, geese, and ducks, and a number of sea bird species are all eaten (Eberhardt, 1976; Fay and Stephenson, 1989; Garrott, 1980; Garrott *et al.*, 1983; Hersteinsson and Macdonald, 1982; Hersteinsson *et al.*, 1989; Novikov, 1962; Chesemore, 1968b, 1975; Banfield, 1974). Carrion figures prominently in the diet. Arctic foxes eat carrion from along the shoreline, and wherever else they can find it. They follow polar bears or wolves and feed on the remains of their kills; they also eat reindeer and seal carcasses (Andriashek *et al.*, 1985; Garrott *et al.*, 1983; Banfield, 1974; Chesemore, 1975; Novikov, 1962; Hersteinsson and Macdonald, 1982; Stroganov, 1962). Seal pups are important prey. It is unlikely that arctic foxes kill seals larger than young pups. In Iceland they may prey upon sheep and lambs, but on the whole these are relatively unimportant food sources (Hersteinsson *et al.*, 1989; Hersteinsson and Macdonald, 1982; Banfield, 1974). Along coastlines, arctic foxes consume various littoral animals, such as sea urchins, crabs, and all sorts of marine mollusks, as well as fish trapped in tidal pools or shallow water, along with other edible flotsam. They also eat insects, including maggots (West, 1987; Banfield, 1974; Chesemore, 1975; MacPherson, 1969; Novikov, 1962; Stroganov, 1962; Hersteinsson and Macdonald, 1982).

Vegetable material consumed includes berries, grasses, various herbaceous plants, seaweeds, and algae. Garrott *et al.* (1983) recorded that vegetable material was found in 79% of all scat samples collected, but it was usually found in trace amounts and perhaps had been ingested accidentally. Arctic foxes are attracted to human habitation, where they obtain food directly from humans and from garbage dumps. Eberhardt (1976) stated that adult foxes teach their young to obtain food in this way. Arctic foxes are sometimes cannibalistic; adults may kill and eat other adults or pups (Banfield, 1974; Chesemore, 1975). The specific conditions that give rise to this behavior are unknown.

Caching is a highly developed behavior and plays an important part in resource management: Food cached at one time may be a crucial supplement for a later, leaner time. Caches can be very large (Banfield, 1974; Ewer, 1973; Chesemore, 1975). See Fay and Stephenson (1989) and West (1987) for recent, detailed diet studies.

ACTIVITY

During the brief arctic summer, when the sun remains above the horizon for most of the day and night, adult arctic foxes retain their nocturnal activity patterns. During the nonsummer seasons they are chiefly, but by no means exclusively, nocturnal (Underwood, 1983; Novikov, 1962; Garrott and Eberhardt, 1987; Banfield, 1974).

REPRODUCTION

Pairs mate from February to May; the timing of breeding varies from region to region. Females reach sexual maturity at 9–10 months, and thereafter they are annually monestrous. [Stroganov, (1962) stated that sexual maturity is attained in the second year of life.] During mating season, breeding pairs spend a great deal of time together, and the females behave submissively toward the males. Proestrus is not accompanied by any overt physical signs, i.e., there is no visible vaginal bleeding. Gestation is usually 51 or 52 days (range 49–57 days). Kits are born in April, May, June, or July (Kleiman, 1968; Hersteinsson and Macdonald, 1982; Chesemore, 1975; Garrott and Eberhardt, 1987; Banfield, 1974; Novikov, 1962; Stroganov, 1962).

Average litter size covaries with the prey base; in years when food resources, particularly lemming populations, are low, litter size is usually three to six. In other years it may be six to nine. In some regions arctic fox populations cycle every 3–5 years. Again, these population cycles are closely tied to the population levels of small rodents. Enormous fluctuations in the number of breeding pairs occur in Sweden, and are linked to lemming population cycles (Hersteinsson, 1984). In

Iceland, there are less radical population responses to ptarmigan 10-year cycles (Hersteinsson *et al.*, 1989).

Arctic foxes seem to be monogamous and to mate for life (Chesemore, 1975; Banfield, 1974; Hersteinsson and Macdonald, 1982; Stroganov, 1962). Novikov (1962) recorded that under natural conditions foxes are monogamous, but in captivity polygyny may occur. Hersteinsson and Macdonald (1982, citing Boitzov, 1937) reported that under the seminatural conditions of island fur farms, some males will mate with more than one female. A 1 : 1 male : female sex ratio seems to be usual (Hiruki and Stirling, 1989). Males of mated pairs assist in kit rearing, and provision and guard the kits and their mate (Banfield, 1974; Stroganov, 1962; Pedersen, 1975; Garrott, 1980; Ewer, 1973). Garrott (1980) stated that interactions between adults and kits are generally limited to the transfer of food. Arctic foxes may use two dens simultaneously, and litters may be split between more than one den (Eberhardt *et al.*, 1983b; Eberhardt, 1976).

In a free-ranging Icelandic population, nonbreeding yearling females assisted mated pairs (Hersteinsson and Macdonald, 1982). These nonbreeders behaved submissively toward both members of the breeding pair, and they all occupied the same range. These females acted as helpers at the den for the first 6–8 weeks of the kits' life, then dispersed. Those tracked a year later had emigrated and had litters of their own. "The morphological characteristics of non-breeding females strongly suggested that they were relatives, presumably offspring of the mated pair" (Hersteinsson and Macdonald, 1982, p. 279).

Litters may stay together for a few weeks or months and then disperse (Underwood, 1983). Chesemore (1975) stated that in Alaska the young of the year disperse in mid-August, and Stroganov (1962) stated that dispersal occurs in autumn. Eberhardt *et al.* (1983b), who radio-tracked 35 arctic foxes in northern Alaska, recorded that dispersal occurred in both fall and late winter/early spring. Most juveniles remained in their natal home ranges through January, and thus relatively few dispersed in the fall. Schwarz (1966, cited in Northcott, 1975) has shown by tag returns that arctic foxes may disperse several hundred kilometers from their natal territory. Eberhardt *et al.* (1983b) record that two dispersing males moved distances of 781 and 2,000 km.

Lifespan is usually given as 8–10 years (Banfield, 1974; Novikov, 1962; Stroganov, 1962), but this seems to be an optimistic assessment in comparison with other comparable canid species. This estimate probably represents the maximum, rather than the average, lifespan. Maximum longevity in a population in the Northwest Territories, Canada, was 6 years for males, 7 for females (Hiruki and Stirling, 1989). A captive lived for 15 years (M. Jones, personal communication, cited in Nowak and Paradiso, 1983). Arctic foxes are preyed on by wolves, wolverines, polar bears, snowy owls, large hawks, golden eagles, and jaegers. Red foxes may harry or kill them, too. The single most significant predator overall is man (Chesemore, 1975; Garrott and Eberhardt, 1982; Banfield, 1974; Garrott, 1980). Rabies is endemic in arctic fox populations, and is periodically responsible for extensive mortality (Garrott and Eberhardt, 1987).

SOCIAL ORGANIZATION AND BEHAVIOR

Many sources state that arctic foxes are solitary except during the breeding season, when they are monogamously paired. It is probably more accurate to say that the social organization of these foxes, as for many other species of the Canidae, is flexible. Using radiotracking and direct observation on the northwest coast of Iceland, Hersteinsson and Macdonald (1982) have conducted the most detailed research to date on this subject. They found a social organization that differs from the commonly accepted structure of solitary and monogamously mated pairs. In their study area, groups were composed of one adult male and two adult females who lived all together with the young of the year. One of the adult females was subordinate to both members of the mated pair; thus there appeared to be dominance hierarchies within these groups. These subordinate females did not breed but acted as helpers, feeding the young of the dominant females who were often their kin; they dispersed to raise their own litters in the following year. Garrott (1980) stated that social behavior within family groups most often occurred between litter mates. Adults spent relatively little time at the den site. On the whole, interactions among the members of a family group were limited. Outside the breeding and pup-rearing seasons, little is known about the social organization of arctic foxes. They appear to make a transition to a transient solitary status, until the following breeding season (Garrott and Eberhardt, 1987).

There are few data on the home range sizes of arctic foxes, and little information on territoriality and its effect on distribution. Arctic foxes do maintain territories (Hersteinsson and Macdonald, 1982; Eberhardt *et al.*, 1983b). In northern Alaska territorial defense seems to be weaker during fall and winter (Eberhardt *et al.*, 1983b). The home ranges of groups in Iceland overlap very little, if at all, with those of adjacent groups (Hersteinsson and Macdonald, 1982). On the northwest coast of Iceland, home ranges vary in size from 8.6 to 18.5 km^2, although after a population crash the density may be only 0.086 foxes per km^2.

Hersteinsson and Macdonald's (1982, p. 277) observations suggested that intruders in neighboring territories are sometimes attacked and expelled by residents. Consequently "these group ranges constituted territories from which their occupants rarely strayed." During the 2 years of their study, the territorial boundaries of the groups they observed remained very similar. Urine marking and barking function in territory maintenance, occurring along territory boundaries. Both barking and urine-marking behaviors are performed by animals of both sexes. Arctic foxes have been observed actively defending their territory, driving intruders away (Chesemore, 1967).

Den distribution is highly variable. In prime habitat dens may be located extremely close together, while elsewhere they are much farther apart. In favored habitat, i.e., along a river valley, dens occur at a density of 2 per km^2. In rare instances burrows are only 0.5–1 km apart. In the USSR there are 1–6 dens per 10 km^2, and in the Northwest Territories there is 1 den per 36 km^2 (Chesemore, 1975). Banfield (1974) stated that on average dens are 900 m apart. When populations are peaking, there may be as many as six dens per km^2. Eberhardt *et al.* (1983b) gave a figure of one den per 12 km^2 on Prudhoe Bay, and one per 34 km^2 on the Colville River Delta. Novikov (1962) stated that even where arctic foxes are numerous, the distance between burrows is 3–10 km and sometimes 30 km or more.

Individuals congregate at abundant food sources. Groups of up to 40 gather at large carcasses. These groups fight among themselves (Underwood, 1983; Chesemore, 1975; Banfield, 1974).

Growls, barks, and coughs are common vocalizations. Coos occur in greeting or contact-seeking contexts (Cohen and Fox, 1976). Screams and hissing accompany intraspecific agonistic interactions (Banfield, 1974). Hersteinsson and Macdonald (1982) stated that the vocal (and olfactory) behaviors of *Vulpes* and *Alopex* appear to be quite similar. Arctic foxes also employ a rich and varied repertoire of visual signals for intraspecific communication (Wakely and Mallory, 1988).

Small-Eared Dog *(Atelocynus microtis)*
Credit: Thomas McHugh/National Audubon Society
Collection/Photo Researchers, Inc.

CHAPTER 2
Genus *Atelocynus*

Atelocynus microtis: Small-Eared Dog

Known as the small-eared dog, small-eared zorro (fox), or zorro negro, *Atelocynus microtis* is remarkable in both appearance and distribution. It is one of the rarest and least studied wild canids, and nothing is known about its behavior in the wild (Langguth, 1975b). The scant behavioral information available has been collected from observations of captive animals.

DISTRIBUTION AND HABITAT

Small-eared dogs are found in the Amazon, Upper Panama, and Orinoco basins in Colombia, Ecuador, Brazil, Peru, and Venezuela. They live in tropical forest habitat from sea level to 1,000 m (Berta, 1986; Ewer, 1973; Nowak and Paradiso, 1983). Langguth (1975a) stated that all specimens collected have been found in lowland tropical rainforest, making this species the only member of the Canidae that exists in this type of habitat. Distribution maps can be found in Berta (1986), Hershkovitz (1961), and Stains (1975).

PHYSICAL CHARACTERISTICS

The physical appearance of small-eared dogs deviates markedly from the typical canid pattern. The ears are short, rounded, and smaller relative to body size than is usual for canids. The body is stocky and compact, and the legs are very short. There is a superficial resemblance to the bush dog, *Speothos venaticus*. Body length is 72–100 cm, tail length is 25–35 cm, and height at shoulder is about 35 cm. Weight is roughly 9 kg. The pelage is short, sleek, and grizzled dark brown to iron gray in color. There is a dark band along the spine, extending down the tail. Limbs and tail are nearly black, and the underbelly is dark, except for the pelvic region where the hair is lighter. The ears are a rufous color both internally and externally. The tail is bushy, and there may be a round patch of white beneath its root (Hershkovitz, 1961; Mivart, 1890; Berta, 1986; Clutton-Brock *et al.*, 1976; Nowak and Paradiso, 1983; Stains, 1975).

The teeth are heavy and longer than those of other South American canids (with the exception of the maned wolf, *Chrysocyon brachyurus*). The canines are long and "fox-like," and the cheek teeth are robust, with a much enlarged second molar (Clutton-Brock *et al.*, 1976; Berta, 1986; Stains, 1975).

TAXONOMY

The taxonomic positions of the South American canid species have been unclear from the outset, and the position of the small-eared dog is no exception. Once placed in the genus *Cerdocyon* (Thomas, 1914, cited in Clutton-Brock *et al.*, 1976), it has since been put in the subgenus *Dusicyon*, and then into its own genus, *Atelocynus* Cabrera, 1940. Simpson (1945) acknowledged, but did not follow, this classification. Clutton-Brock *et al.* (1976) placed the small-eared dog in the genus *Dusicyon*. Their analysis showed similarity with the crab-eating fox (*Cerdocyon thous*) (which they classified as *Dusicyon thous*). According to their analysis, small-eared dogs lie at the periphery of the *Dusicyon* group. Recent analysis shows that *Atelocynus* is closely related to *Speothos venaticus*, the bush dog (Berta, 1986). Berta's (1987) cladistic analysis placed *Atelocynus* within the *Cerdocyon* group, along with *Cerdocyon*, *Speothos*, and *Nyctereutes*.

Langguth (1975a) granted *A. microtis* full generic rank, as did Nowak and Paradiso (1983) and Hershkovitz (1961). Van Gelder (1978), in keeping with his radical revision of the taxonomy of the Canidae, placed the species within the genus *Canis*, as subgenus *Atelocynus*, *Canis (Atelocynus) microtis* Sclater, 1882.

On the whole, it seems that the skeletal and overall physical characteristics of small-eared dogs are singular enough to warrant monotypic generic status. However, resolution of these taxonomic disagreements awaits further research on members of this little-studied species. Until that time, the most appropriate taxonomic designation is in the monotypic genus *Atelocynus* Cabrera, 1940, as *Atelocynus microtis* (Sclater, 1882) (Berta, 1986, 1987). There are no subspecies (Berta, 1986; Stains,

1975). No fossils are known (Berta, 1987). See Wurster and Benirschke (1968) for a karyotype (2n = 74–76: NF = 76).

DIET

The food habits of free-ranging small-eared dogs are unknown (Langguth, 1975a). A captive pair ate raw meat, occasional shoots of grass, and the "common food of the people," whatever those are (Hershkovitz, 1961, p. 507).

ACTIVITY

No information on activity patterns is available. The small-eared dog is probably nocturnal.

REPRODUCTION

No information on reproduction is available (Berta, 1986).

SOCIAL ORGANIZATION AND BEHAVIOR

Hershkovitz (1961) observed two captives, one of each sex, at the Brookfield Zoo, Chicago. He stated that these may have been the first pair of small-eared dogs to be exhibited as live zoo specimens. The following information is drawn entirely from his account of their behavior.

The male of the pair was docile, friendly, and quite playful. He recognized familiar humans and "very obviously enjoyed being the object of attention." He was, overall, quite tame and allowed himself to be patted by familiar humans. In contrast, the female was distinctly hostile toward all humans and growled at them. As for interactions between the two animals, the male, though smaller, dominated the female in most activities, and displaced her at their common feeding dish. Sometimes, however, the dominance of the male at the food dish was reversed for a brief time. No biting or fighting between the two was observed, and the pair occupied a common sleeping box.

Coyote *(Canis latrans)*
Credit: Henry Holdsworth

CHAPTER 3
Genus *Canis*

Canis adustus: Side-Striped Jackal

In comparison with its congeners the black-backed and golden jackals (*C. mesomelas* and *C. aureus*), side-striped jackals are relatively rarely observed in the field (van Lawick-Goodall and van Lawick-Goodall, 1971; Bekoff, 1975). Consequently, comparatively little is known about their natural history.

DISTRIBUTION AND HABITAT

Members of this species occur from northern South Africa northward to Ethiopia on the east, and westward through Nigeria up to Upper Gambia on the west (Rosevear, 1974; Kingdon, 1977; Müller-Using, 1975d). They are found in the Ngorongoro Crater basin and in. the Serengeti region, but have been only rarely observed there during the better known long-term field studies of the carnivores of these areas (van Lawick-Goodall and van Lawick-Goodall, 1971; Schaller, 1972; Kruuk, 1972a).

The relative scarcity of observations of these jackals is related to the fact that they prefer wooded and densely vegetated habitat (Kingdon, 1977; van der Merwe, 1953a; Clutton-Brock *et al.*, 1976), habitats that make observation more difficult than on the open plains where the other jackal species occur. Side-striped jackals also frequent cultivated areas. They are found in the mountains up to 2,700 m (Kingdon, 1977). This is the most widespread of the four jackal species in all moist environments, and may be found in swampy areas as well (Kingdon, 1977). In some regions side-striped, black-backed, and golden jackals are sympatric, but side-striped jackals generally prefer moister areas than golden jackals do (Rosevear, 1974), and habitats with denser vegetation than black-backed jackals prefer (Clutton-Brock *et al.*, 1976). In Kenya, in an area of sympatry, *C. adustus* was located more often

23

in open *Euphorbia* woodland, *C. mesomelas* in closed woodland, and *C. aureus* in grassland habitat (Fuller *et al.*, 1989). In this way, habitats and resources within a single region are partitioned among the three similar species. See Fuller *et al.* (1989) for a discussion of ecological segregation in an area of sympatry.

PHYSICAL CHARACTERISTICS

Head-plus-body length of side-striped jackals varies from 65 to 81 cm. Height at shoulder is 41–50 cm. The tail length is 30–41 cm. Weight ranges from 6.5 to 14 kg (Kingdon, 1977). *Adustus* means sunburned or swarthy. The general color is grayish to fawn, with whitish underparts and lighter sides (Dorst and Dandelot, 1969). Along the sides there is "an ill-defined light colored or whitish line from shoulder to root of tail, bordered sometimes with black at its lower margin" (Dorst and Dandelot, 1969, p. 91). This stripe of guard hairs running along each side of the body is what gives this animal its common name. The tail is bushy and darker than the body. There is usually a conspicuous white tip to the tail, though this may not be present on North African specimens where the tail tip is black instead (Kingdon, 1977; Dorst and Dandelot, 1969).

Taller at the shoulder than black-backed jackals (*C. mesomelas*), side-striped jackals are also larger, stockier and more heavily built (Shortridge, 1934; van der Merwe, 1953a,b; Clutton-Brock *et al.*, 1976). Note that these three jackal species have a high degree of intraspecies size variation and some interspecies size overlap. Side-striped jackals have comparatively shorter ears and a blunter, more wolfish muzzle than those of other jackal species (Kingdon, 1977; Dorst and Dandelot, 1969).

The skull is longer and narrower than those of black-backed jackals. The teeth are smaller; the carnassials are particularly diminished relative to those of golden and black-backed jackals (Rosevear, 1974; Clutton-Brock *et al.*, 1976). The molars, however, are relatively larger (Ewer, 1973). The mandible is less powerfully built than in the golden jackal (Rosevear, 1974). Skull and dentition characteristics confirm the observation that side-striped jackals take smaller prey than do black-backed jackals (Clutton-Brock *et al.*, 1976). The dental formula conforms to the usual canid pattern: incisors 3/3, canines 1/1, premolars 4/4, molars 2/3 = 42.

TAXONOMY

Side-striped jackals are highly similar and closely related to black-backed jackals (*C. mesomelas*) (Clutton-Brock *et al.*, 1976). The species is monotypic (Stains, 1975). See Wayne *et al.* (1989) for a discussion of sympatry and morphological divergence in the side-striped, golden, and black-backed jackals (along with a number of other canid species). They found minimal dental, cranial, and size divergence, despite apparent divergence of the three species at least 2 million years ago (based on mtDNA restriction-site polymorphisms). As is also the case for all other members of the genus *Canis*, the karyotype has 2n = 78: NF = 80 (Wurster and Benirschke, 1968).

DIET

Side-striped jackals are thoroughly omnivorous, more so than the other jackal species. Their diet includes carrion, insects, fruit and vegetable material, and small vertebrates such as birds, mice, and reptiles (Kingdon, 1977; Bothma, 1971a; van der Merwe, 1953a; Dorst and Dandelot, 1969; Shortridge, 1934). Banana peels and rice are eaten by captives, which indicates the lack of selectivity in diet (Bothma, 1971a; Müller-Using, 1975d). Side-striped jackals may consume domestic stock in the form of carrion, but there is general agreement that they pose no threat to livestock (Kingdon, 1977).

ACTIVITY

These jackals are primarily nocturnal, particularly in cultivated areas, but they may be active during morning or evening hours as well (Kingdon, 1977; Rosevear, 1974; Shortridge, 1934; Fuller *et al.*, 1989). During cooler weather, they are active on the veldt in daylight hours (van der Merwe, 1953a; Shortridge, 1934). Side-striped jackals are seen less often than black-backed jackals both because they are more wary and because they prefer areas with dense ground cover (Rosevear, 1974; Dorst and Dandelot, 1969).

REPRODUCTION

Gestation is reported as between 8 and 10 weeks (Dorst and Dandelot, 1969; van der Merwe, 1953a; Rosevear, 1974; Shortridge, 1934; Kingdon,

1977). Litter size is usually three or four, but can be as many as seven (Rosevear, 1974; Kingdon, 1977; van der Merwe, 1953a). Shortridge (1934) stated that interbreeding with domestic dogs occurs.

SOCIAL ORGANIZATION AND BEHAVIOR

This is a solitary to moderately social species. The essential social unit is the mated pair plus offspring, but side-striped jackals are often seen singly as well. As many as 12 individuals may gather at a kill (Kingdon, 1977). Van der Merwe (1953a) and Sclater (n.d., cited in Shortridge, 1934) stated that side-striped jackals frequently hunt in packs, but no recent observations have substantiated this. Side-striped jackals usually forage singly, in pairs, or in parent-offspring groups. When a concentrated food source is present, larger numbers may gather to exploit the resource (Kingdon, 1977). Kingdon (1977, p. 26) observed that in southern Uganda, where they are common, side-striped jackal pairs are "well spaced out." A single side-striped jackal in Kenya had a minimum home range of 1.1 km² (Fuller et al., 1989).

Short-range vocalizations include whines and a low chattering sound; both have been observed in distress contexts. Yaps, yelps, and barks occur. Long-range vocalizations include yapping howls and hoots, but unlike golden and black-backed jackals, side-striped jackals do not howl.

Canis aureus: Golden Jackal

This is one of four species of jackals that occur in Africa (C. aureus, C. mesomelas, C. adustus, C. simensis). In terms of physical and behavioral characteristics, golden jackals vary considerably from region to region, but they are basically medium-sized, moderately social canids. Common names include Asiatic, golden, and common jackal.

DISTRIBUTION AND HABITAT

Golden jackals occur in northern Africa and across a wide swath of Eurasia. Specifically, their range extends from Senegal on the northwestern coast of Africa, north to Morocco and the Mediterranean region, through the northeastern part of the Congo, and eastward to

Kenya. They are found across southeastern Europe and the Arabian Peninsula. In southern Asia they occur throughout the Indian subcontinent and Sri Lanka, and as far east as Burma and Thailand (Rosevear, 1974; Kingdon, 1977). Recent range expansion has resulted in documented sightings as far northeast as Austria (Hoi-Leitner and Kraus, 1989). Populations have become so firmly established in Bulgaria that they are no longer protected and are actively controlled (Genov and Wassilev, 1989). Golden jackals have the northernmost distribution of the four jackal species.

Golden jackals inhabit both wooded and open country from sea level to over 1,000 m (Clutton-Brock *et al.*, 1976; Müller-Using, 1975d). They are found in the Arabian and Sahara deserts, and in India in a variety of habitats, as well as on the open plains and grasslands of northern and eastern Africa (Kingdon, 1977). In parts of their range they have adapted to the presence of man, and may enter cities and villages, but only at night (Dorst and Dandelot, 1969).

PHYSICAL CHARACTERISTICS

Golden jackals are medium-sized canids, with a relatively unspecialized physical appearance markedly similar to that of coyotes (*C. latrans*). Head-plus-body length ranges from 60 to 106 cm. Height at shoulder is 38–50 cm. Tail length is 20–30 cm. Weight ranges from 7 to 15 kg (Kingdon, 1977). In Bangladesh, average male weight is 10.3 kg, and that of females averages 8.5 kg (n = 17) (Poché *et al.*, 1987). As the specific Linnaean name indicates, the basic pelage color is gold. It varies from sandy gold to yellow with a reddish tinge through tawny to rufous. Individuals dwelling in mountainous and rocky areas are grayer and in winter have darker tips to the guard hairs. Underparts are pale ginger to cream or nearly white. The tail tip is black. Occasional melanistic individuals have been reported (Dorst and Dandelot, 1969; Kingdon, 1977; Müller-Using, 1975d). Females have eight mammae.

The skull is like that of a very small wolf and has well-developed, high-crowned teeth. Surprisingly, it is more similar to skulls of coyotes, dingoes, and wolves than to those of black-backed, side-striped, and Ethiopian jackals (Clutton-Brock *et al.*, 1976). Wayne *et al.* (1989) found that golden, black-backed, and side-striped jackals are strikingly similar morphologically, to a much greater degree than other sympatric canids (i.e., South American foxes). This is true despite apparent long-term sympatry, which flies in the face of current ideas of competitive exclusion. The similarities may be explained in part by resource partitioning (Fuller *et al.*, 1989), as well as by the relative diversity of prey and predator species in Africa (Wayne *et al.*, 1989).

TAXONOMY

Based on analysis of skull, skeleton, internal anatomy, body proportions, and behavioral characteristics, Clutton-Brock *et al.* (1976) considered *C. aureus* to be the most typical species of the genus *Canis* Linnaeus, 1758. About 12 subspecies are recognized (Stains, 1975). See Rosevear (1974) for a discussion of unresolved taxonomic quibbles. See Ferguson (1981) for a discussion of *C. aureus* and *C. lupus* taxonomy in North Africa and the Middle East. The karyotype has 2n = 78: NF = 80 (Wurster and Benirschke, 1968).

DIET

Golden jackals, like coyotes, are unspecialized, opportunistic, and flexible with respect to diet. Feeding habits vary across their range. They are capable hunters (van Lawick-Goodall and van Lawick-Goodall, 1971) and frequently kill full-grown as well as juvenile antelopes of the smaller species, such as duikers or gazelles. In order to capture prey of this size, three or more jackals participate in the hunt (Rosevear, 1974). Two instances of cooperative hunting by golden jackals on langurs (*Presbytis pileata*) have been described by Stanford (1989). Predation and predation attempts by golden jackals on langurs (*Presbytis entellus*) in India are described in Newton (1985). Groups may kill sheep and goats (McShane and Grettenberger, 1984). Golden jackals eat small to medium-sized vertebrates, including ground birds and their eggs, rodents, lizards, and snakes. They also prey upon domestic fowl (Poché *et al.*, 1987), and consume all sorts of plant material and many types of insects (McShane and Grettenberger, 1984; Poché *et al.*, 1987; Kingdon, 1977). Carrion is an important food source in some regions (Poché *et al.*, 1987).

Wyman (1967) described how black-backed jackals and golden jackals partition food resources in areas where their distributions overlap, and discussed the diets of both species in detail. See Fuller *et al.* (1989)

for other information on resource partitioning among sympatric jackal species.

ACTIVITY

Activity patterns are highly variable. These jackals are strictly nocturnal in areas close to human habitation (Kingdon, 1977), though they may be at least partially diurnal in other areas (van Lawick-Goodall and van Lawick-Goodall, 1971; Dorst and Dandelot, 1969). Wandrey (1975) reported consistent crepuscular activity patterns in the captive jackals he observed, noting that the artificial feeding schedule may have influenced them.

REPRODUCTION

Females are annually monestrous. Gestation is about 63 days. Litter size ranges from one to nine; two to four is usual. Wyman (1967) observed that litters in the Serengeti had an average of only two pups. Van Lawick-Goodall and van Lawick-Goodall (1971) reported an instance of two litters born to the same female in one year. This is an unusual but not unique ability among the Canidae. Males participate in raising the pups, and offspring from previous years may participate as well (Moehlman, 1983). Life span is up to 18 years in captivity (Müller-Using, 1975d).

SOCIAL ORGANIZATION AND BEHAVIOR

Recent field studies give a picture of a moderately social canid (Moehlman, 1983; Macdonald, 1979b; Golani and Keller, 1975; van Lawick-Goodall and van Lawick-Goodall, 1971). As is also the case for many of the less specialized canids, the social system is strikingly flexible. Its form seems to reflect the abundance and distribution of resources, particularly food. The basic social unit is the mated pair and its offspring, which may include, in addition to the current pup crop, offspring of a previous year that have not yet dispersed. Larger packs are probably family groups, though associations of unrelated individuals also occur (Golani and Keller, 1975). On a nature reserve in Israel, Macdonald (1979b) observed much larger social groups of 10–20 golden jackals. These groups had stable compositions and fixed home ranges. Macdonald interpreted this arrangement as an adaptation to highly clumped and defensible food resources (the jackals in this study area were provisioned by humans once weekly).

Moehlman (1983), in a study of free-ranging golden jackals near Lake Ndutu on the Serengeti, described a fairly high degree of sociality characterized by long-term pair bonds, cooperative hunting, food sharing, and the maintenance of territories year-round. She observed offspring from previous years helping in the provisioning, guarding, and socialization of new litters of their siblings. Of the eight litters she observed, 100% of the surviving young stayed with their parents through their first year, although individuals also left their natal territories for extended periods before returning. Affiliative behaviors, such as mutual grooming and a greeting ceremony, occurred commonly among the golden jackals observed by van Lawick-Goodall and van Lawick-Goodall (1971).

Mated pairs are territorial. They mark and vigorously defend areas around their dens. Boundaries are demarcated by urination and defecation (Rosevear, 1974). It appears that hunting ranges are much more loosely defined, and defense of these against other jackals is sporadic and somewhat inconsistent (van Lawick-Goodall and van Lawick-Goodall, 1971). The area of these hunting ranges varies considerably depending on environmental conditions, from 2.5 to 20 km². In an agricultural area of Bangladesh with heavy human habitation, a male had a home range of 1.1 km², a female 0.6 km² (Poché et al., 1987). Outside of the breeding season, some individuals may wander long distances (Kingdon, 1977). Both Moehlman (1983) and Macdonald (1979b) provided detailed discussions of social organization in golden jackals.

Canis latrans: Coyote

On the North American continent coyotes are now the most widespread wild members of the family Canidae. They are highly adaptable and exist in an enormous variety of habitats. Because of their importance as predators on domestic stock, and their furbearer status, a great deal of research has been conducted on the species. More is known about coyotes than about any other canids except wolves and red foxes. The common name is derived from the Aztec word *coyotl*, which means "barking dog."

DISTRIBUTION AND HABITAT

Coyotes exist in North and Central America. Over the past several hundred years their range has expanded outward from a distribution centered in the western United States (Nowak, 1978). The range of this species currently encompasses most of the continental United States and Canada southward to Costa Rica and Panama. Coyotes are found from northern Alaska eastward to Hudson's Bay, Quebec, New Brunswick, and Nova Scotia. The extirpation of competitors and natural enemies, as well as the growth of agricultural and land-clearing practices, seems to have been responsible for this range expansion. Escapes of captives and purposeful release of animals intended to provide recreational hunting opportunities have also contributed to range expansion (Hill *et al.*, 1987). Despite long-term attempts by humans to reduce their population by poisoning, trapping, shooting, and denning, this expansion is continuing (Gier, 1975).

Originally grassland animals, coyotes now exist in open forests, forest-edge habitat, deserts, and agricultural and urban areas. They are quite tolerant of human activities and human-wrought habitat changes (see Andelt and Mahan, 1980), characteristics that have contributed to their recent success. Dens, either dug by coyotes themselves or modified from the existing excavations of other animals, such as badgers, marmots, and skunks, are located in brushy areas, among rock ledges, in embankments, or in open country. Like wolves, a mated pair of coyotes may return to the same den site year after year (Bekoff and Wells, 1980). See Voigt and Berg (1987) for a detailed distribution map.

PHYSICAL CHARACTERISTICS

Coyotes resemble wolves, sometimes to such a degree that they are mistaken for them. Coyotes are smaller (although there is some size overlap), with a slenderer build, proportionately larger ears, and slender legs. Size varies regionally and between subspecies, and males are larger and heavier than females. Body length is 100–140 cm, with a tail length of 30–40 cm. Weight ranges from 7 to 20 kg: Adult males weigh

8–20 kg, adult females 7–18 kg (Banfield, 1974; Bekoff, 1977; Gier, 1975). The largest coyote on record weighed 34 kg (Young and Jackson, 1951). Northern and mountain populations are larger and more heavily furred than those from southern and desert habitats. The largest coyotes are found in northeastern North America. Individuals from Mexican deserts average 11.5 kg; those from Alaska, 18 kg. Banfield (1974) gave an average weight of 13.2 kg for northern coyotes.

Coyotes are among the fastest terrestrial mammals in North America, able to attain speeds of up to 64 kph (Zeveloff, 1988).

Pelage color includes shades of buff, brown, gray, rufous, and blends of these colors. Long, banded, black-tipped guard hairs give a grizzled aspect to the fur. Flanks are grizzled or fulvous. Underfur is cream, gray, or fulvous. At higher latitudes, pelage tends toward gray or black tones, while in deserts it is more fulvous in color. Underparts are pale cream to white, and the throat is paler as well. Tails are well furred, less than half the body's length, and black-tipped. A distinct black spot overlies the caudal gland. Ears are large, and pale inside. The muzzle is slender and fulvous, the chin pale. There is a distinctive dark line running vertically down the inside of the foreleg.

Skull size is intermediate between those of wolves and of jackals. The braincase is proportionately larger than in wolves (Clutton-Brock et al., 1976; Bekoff, 1977). The teeth are well developed, the canines relatively long and narrow (Clutton-Brock et al., 1976; Hall and Kelson, 1959). The dental formula is incisors 3/3, canines 1/1, premolars 4/4, molars 2/3 = 42. The sagittal crest is more developed in males than in females (Voigt and Berg, 1987).

TAXONOMY

In terms of physical characteristics, coyotes lie between wolves and jackals (Clutton-Brock et al., 1976). See Bekoff (1977) for a summary of these differences. See Lawrence and Bossert (1975) for multiple character analysis of these species. Coyotes were present as a species distinct from C. lupus during the Pleistocene (Gier, 1975): C. latrans and C. lupus probably diverged from a common ancestor in the late Pliocene, 2–3 million years ago (Wayne and O'Brien, 1987). Coyotes, wolves, jackals, and domestic dogs can all mate successfully with each other (Bekoff, 1977; Kennelly and Roberts, 1969), but behavioral and physical barriers generally constrain interbreeding beyond the first generations in free-ranging conditions. F1 coyote x beagle hybrid males produce sperm year-round, in contrast with the seasonal spermatogenesis of pure *C. latrans* (Kennelly and Roberts, 1969). Extensive, transgenerational interbreeding between coyotes and wolves in some parts of North America may have occurred. Karyotype has 2n = 78, the same as for wolves, domestic dogs, and golden jackals (*C. aureus*) (Wurster and Benirschke, 1968).

Nineteen subspecies were recognized by Young and Jackson (1951). Bekoff (1977) also lists 19, although later commenting (Bekoff and Wells, 1986) that due to movement patterns and interbreeding this subspecific classification is now of limited value. Nowak (1978) also commented that this classification scheme is of dubious validity.

DIET

Coyotes are opportunistic predators and scavengers. In addition to vertebrate prey, vegetable material and invertebrates figure prominently in their diet. In the northerly parts of their distribution there are marked seasonal variations in diet. Large ungulates (often in the form of carrion) predominate in winter, rodents or lagomorphs in summer (Parker and Maxwell, 1989; Bowen, 1978; Bekoff and Wells, 1980). Elk, moose, bison, and deer are all consumed. In the eastern United States, white-tailed deer predominate (Harrison and Harrison, 1984; MacCracken and Uresk, 1984; Drewek, 1980; Bekoff and Wells, 1980; Bowen, 1978; Gier, 1975). In Jasper, Alberta, in winter, carrion of large ungulates, primarily elk, constituted 67% of the diet; in summer, ungulates composed 50% of the diet (Bowen, 1978). Coyotes generally hunt singly, although they sometimes engage in cooperative hunting of larger prey (Rathbun *et al.*, 1980; Hamlin and Schweitzer, 1979).

Rodents, lagomorphs, and other small mammals are important in coyotes' diet. During nonwinter seasons, coyotes on the National Elk Refuge, Wyoming, relied primarily on small rodents (Bekoff and Wells, 1980). In winter in Jasper, these constituted 23% of the diet (Bowen, 1978). In Maine during May through October, snowshoe hare occurred in 30% of scats, small mammals in 21% (Harrison and Harrison, 1984). Small mammals constituted 49.9% by dry weight of scats from Arizona (Drewek, 1980). In South Dakota rabbits constituted 16.3% of scats by dry weight; *Microtus* spp. were 12.2% (MacCracken and Uresk, 1984).

Amphibians, reptiles, and fish are all consumed, in relatively small quantities. Coyotes also eat invertebrates, including crustaceans and insects, and occasionally catch birds (MacCracken and Uresk, 1984; Drewek, 1980; Harrison and Harrison, 1984; Elliott and Guetig, 1990). The proportion of vegetable material in the diet varies seasonally, increasing in the summer and fall. Both wild and cultivated fruit are eaten, including blueberries, blackberries, persimmons, prickly pears, apples, peaches, pears, and melons of all sorts. Coyotes also eat peanuts, carrots, cultivated grains, and wild grasses (Gier, 1975; Harrison and Harrison, 1984; Drewek, 1980; Andelt *et al.*, 1987).

Coyotes are infamous for preying on domestic stock, particularly sheep and poultry (Sterner and Shumake, 1978), although it is clear that poor stock-husbandry practices exacerbate the problem. It is sometimes assumed that coyotes adversely affect game populations, but this has not been clearly demonstrated (Clark and Stromberg, 1987). Ungulate fawns and infirm adults are preyed on preferentially (Bowyer, 1987). Coyotes have killed and eaten domestic dogs (Bider and Weil, 1984), and, on extremely rare occasions, they have attacked human children (Carbyn, 1989a). Food items are regularly cached. See Harrington (1981) for a discussion of the "bookkeeping" role of urine marking at emptied cache sites.

ACTIVITY

Coyotes are largely crepuscular and nocturnal. Daylight activity also occurs, particularly in cooler weather and during winter (Drewek, 1980; Banfield, 1974; Bekoff and Wells, 1980).

REPRODUCTION

Females are annually monestrous. [A case of extended or possibly multiple estrus has been observed in a captive female (Harrington and Ryon, 1987).] Some females are capable of breeding in their first year, but the percentage that does so is variable, apparently determined by food abundance, population density, intensity of control measures, and other environmental factors (Kennelly and Johns, 1976; Kennelly, 1978; Bekoff, 1977; Bekoff and Wells, 1986). The percentage of yearlings who do breed fluctuates between 10% and 70% (Gier, 1975).

Large compensatory increases in reproduction rates have been documented in areas of stringent population control (Voigt and Berg, 1987).

Coyotes are generally monogamous. Pair bonds often last for years, though not necessarily for life. In a pack situation typically only a single pair will mate (Camenzind, 1978b; Bekoff and Wells, 1980, 1986). An instance in which two different litters were reared in close proximity (their dens were 35–45 m apart) has been documented (Ortega, 1988): It is possible that two females from the same pack reproduced that year. Courtship activities begin in late December or early January. There is a long proestrous period of 2–3 months, then estrus and mating occur between January and March (Kennelly, 1978; Camenzind, 1978b; Bekoff and Diamond, 1976; Bekoff and Wells, 1986). A copulatory tie 5–20 minutes in length occurs during mating (copulatory ties have been observed in almost all other canid species). Gestation is about 60 days. Litter sizes are influenced by food availability, population density, and level of exploitation. The range is 2–12, with averages between 4 and 7 (Gier, 1975; Camenzind, 1978a; Jean and Bergeron, 1984; Kennelly, 1978; Hall and Kelson, 1959; Bekoff, 1977). A female with 17 pups and 17 placental scars was captured: This is the largest recorded litter (Bowman, 1940, cited in Kennelly, 1978).

The males of mated pairs assist in rearing pups, provisioning the mother and pups, and guarding the den. Offspring from previous litters may act as helpers, participating in territorial defense, den guarding, scent marking, and provisioning of the pups (Bekoff and Wells, 1980, 1986; Gier, 1975; Bowen, 1978). Bekoff and Wells (1986) found that helpers did not play a significant role in provisioning pups and that their presence did not contribute significantly to the pups' survival to 5–6 months of age. The helpers did participate in territorial defense and scent marking.

Pups reach adult size and weight by the end of their ninth month. They may disperse in the fall or winter of their first year, some leaving as early as August while other family groups remain intact through November. Dispersal may often be a gradual process occurring over a period of weeks. Some offspring who do not disperse may become helpers. Others will stay around the edge of their natal territory, interacting with their family only rarely (Bekoff and Wells, 1986; Bekoff, 1977; Gier, 1975). Dispersal is a high risk process, associated with high mortality. Pups fall victim to raptors, domestic dogs, and other coyotes. Disease and parasites take their toll as well. Rabies is uncommon (Voigt and Berg, 1987). Coyotes are killed by humans both intentionally and unintentionally. Large-scale control efforts, including trapping, shooting, poisoning, denning, and killing for pelts, account for both adult and pup mortalities. In one population, 81% of known deaths were caused by humans (Gese *et al.*, 1989). Mortality resulting from coyote-coyote aggression has been recorded (Okoniewski, 1982). See Voigt and Berg (1987) for detailed mortality data. Maximum known longevity in the wild is 15.4 (±1) years: This male reproduced successfully until at least 12.8 years of age (Gese, 1990). A captive lived 21 years 10 months (Nowak and Paradiso, 1983).

SOCIAL ORGANIZATION AND BEHAVIOR

Social organization is highly variable and flexible. The fundamental social unit is the mated pair plus offspring, although coyotes, even those within a single region, may also live as nomads or in stable packs. There are five types of social organization in coyotes: packs, resident pairs, solitary residents, nomads (transients), and aggregations (Camenzind, 1978b; Bowen, 1978; Bekoff and Wells, 1980, 1986).

Packs are groups that occupy and defend a territory, maintain a social hierarchy, and often feed and den together (Camenzind, 1978b, p. 275; Bowen, 1978; Bekoff and Wells, 1986). They are composed of a mated pair and other (usually closely related) individuals, both adults and subadults. A single breeding pair is the nucleus, and there is annual turnover of yearlings and subadults. An unrelated individual may rarely be accepted into a pack (Bowen, 1978). Packs range in size from three to eight. Approximately 61% of the resident coyotes on the National Elk Refuge in Wyoming belong to packs (Camenzind, 1978b). In winter in Alberta, 59% of all coyotes were in packs (Bowen, 1978). Early reports of coyotes stated that they frequently associated in larger groups. Lewis and Clark (cited in Brown, 1973) stated that coyotes generally associated in bands of 10 or 12, and were only rarely seen alone. Dobie (1947, cited in Brown, 1973) recorded that early accounts tell of groups as large as 100.

Resident pairs den together and spend the entire year together. Resident pairs made up 17% of the coyotes studied in Alberta (Bowen,

1978), and 78% of a study population in southeastern Colorado (Gese *et al.*, 1988b). In nonprotected areas this is probably the most common form of social organization. Usually the offspring of these mated pairs disperse in their first year, but some remain with their parents until after the birth of the next litter.

Solitary residents are coyotes that have an established home range with no cohabiting mate or pack (Allen *et al.*, 1987). Nomads are single individuals with no site attachments, who range over large areas (Bekoff and Wells, 1986). They may be subadults, disabled individuals, or occasionally healthy adults. Nomads made up 15% of the population studied in Alberta (Bowen, 1978) and 22% of the study population in southeastern Colorado (Gese *et al.*, 1988b). These lone coyotes are capable of traveling great distances: A radio-collared female traveled 544 km (straight-line distance) in a year (Carbyn and Paquet, 1986).

Aggregations are "ephemeral groups" (Camenzind, 1978b, p. 273). They may be composed of mated pairs, packs, nomads, or winter migrants. On the National Elk Refuge these aggregations occurred near elk carrion, and only during the winter. Aggregations are characterized by their lack of social organization beyond brief dominance-related interactions (Camenzind, 1978b).

The social organization assumed in a particular region seems to be determined at least in part by aspects of food availability, and also by the level of persecution by humans. The nature of the connection between the size of a coyote group and the proportion of large prey in the diet is unclear, although a number of hypotheses have been suggested (Bowen, 1978; Bekoff and Wells, 1980, 1986; Gese *et al.*, 1989). In Grand Teton National Park, Wyoming, winter food availability was a primary determinant of behavior and social organization. Coyotes were found in packs where carrion was "abundant, clumped, and defendable" (Bekoff and Wells, 1986, p. 264). Where carrion was dispersed and scarce, coyotes occurred as mated pairs or loners. Furthermore, data from this Wyoming study suggested that under relatively difficult conditions, such as snowy winters, energetic advantages may accrue to pack-living coyotes over nonpack coyotes. At least in this particular situation, "pack formation in coyotes appears to be an adaptation for the defense of food, rather than the acquisition of live prey" (Bekoff and Wells, 1986, p. 323). In other situations, group size may influence prey selection (Gese *et al.*, 1989). In southern California, larger

groups of coyotes are more likely to pursue deer and to hunt cooperatively than are pairs or single individuals (Bowyer, 1987). In some
regions, group size fluctuates seasonally. It is possible that the increased sociality associated with mating season (mid-winter) may permit successful hunting of large prey (Gese *et al.*, 1989). In sum, it is
not clear whether coyotes assemble in groups for breeding purposes, so
that exploitation of larger prey is simply a secondary effect, or whether
the size of available prey exerts a selective pressure on the size of coyote groups. Increased sociality may be an adaptation to allow more efficient capture and/or defense of ungulate prey (Bowen, 1978; Bowyer,
1987). Of course, these influences need not be mutually exclusive, and
it is likely that a multitude of factors influence group size and strategies of food selection.

There are clear dominance hierarchies within packs. The members
of a mated pair dominate all other pack members (Bekoff and Wells,
1986). Often, neither member of a mated pair dominates the other consistently. Within packs, helpers typically dominate younger coyotes
up to 9 months of age. Territorial and den defense and other activities
are initiated nearly as often by females as by males (Camenzind,
1978b; Bekoff and Wells, 1986; Bowen, 1978). See Bekoff *et al.* (1981)
for observations on agonistic interactions in captive pup litters.

Territoriality is clearly developed in most populations. In Alberta
during winter, and probably throughout the year, packs and pairs defend well-defined territories. In one southeastern Colorado population,
there is no overlap between the home ranges of different resident pairs.
On the National Elk Refuge, packs and pairs are strictly territorial (Camenzind, 1978b). In North Dakota, coyote families occupy contigu-
ous, nonoverlapping territories (Allen *et al.*, 1987). Mated pairs tend to
remain in the same territory. Upon the death of one or both of the
mated pair, territorial boundaries are significantly altered by new residents (Allen *et al.*, 1987). Solitary residents do not defend the home
ranges they occupy (Bowen, 1978).

Home ranges and territories vary from about 1 to 100 km² (Bekoff
and Wells, 1986). Laundré and Keller (1984) and Voigt and Berg (1987)
reviewed home range size in coyotes. See Voigt and Berg (1987) for a review of population density data. There does not seem to be a difference
in the home range sizes of males and females (Bowen, 1978; Bekoff and
Wells, 1980), although while pups are confined to the den area, female

parents' home ranges are smaller (Parker and Maxwell, 1989). Scent marking plays a role in territoriality. Scent marks (both urine and feces) do not serve as barriers against trespassers but rather seem to advertise territorial boundaries, indicating to residents that they are on their own territory and informing trespassers that they are intruding (Bekoff and Wells, 1986). Coyotes do not avoid urine marks of other coyotes from neighboring territories; instead, they vigorously mark over them. Both males and females scent-mark their territories throughout the year (Bowen and Cowan, 1980; Bowen, 1978; Camenzind, 1978b). Howls serve as long-range territorial announcements by broadcasting the location and identity of coyote groups. Howls also perform an intragroup affiliative function. Lehner (1978a,b) has described the vocal repertoire of coyotes, which includes 11 categories, based on acoustic criteria and behavioral contexts.

Aggression and predation by coyotes appear to limit red fox (*Vulpes vulpes*) distribution in some regions (Sargeant and Allen, 1989; Harrison *et al.*, 1989). In turn, wolves seem to displace coyotes. Coyote range expansion may be related to wolf eradication (Harrison *et al.*, 1989; Fuller and Keith, 1981; Krefting, 1969). See Dekker (1989, 1990) for observations on spatial segregation and population fluctuations in sympatric red fox, wolf, and coyote populations in Alberta. It will be interesting to see how proposed wolf reintroductions—for example, in Yellowstone National Park—will affect resident coyote populations.

Canis lupus: Gray Wolf

Wolves once had the largest natural range of any terrestrial mammal except man. Wolf populations have declined steadily since the early 1800s, entirely because of man's efforts. Denning, shooting, trapping, poisoning, aerial hunting, and habitat destruction have all contributed to this decline, and wolf distribution is now a fraction of its former size. The high commercial value of pelts motivates trapping and hunting. Net documented world trade is 6,000–7,000 pelts per year. Wolves are classified as vulnerable by the International Union for the Conservation of Nature and Natural Resources (IUCN), and are on Appendixes I and II of The Convention on International Trade in Endangered Species (CITES) (Ginsberg and Macdonald, 1990). Although wolves have historically had an adversary relationship with humans, there are few documented cases of wolf attacks on humans. At present, the tide of public opinion is turning in the wolf's favor, and there is popular support for the reintroduction of wolves to appropriate protected areas (McNaught, 1987). Given adequate habitat and protection from intense pressure by humans, members of this relatively adaptable species are able to recolonize suitable areas. Proposed reintroduction to Yellowstone National Park will be an interesting test case (see Wilcove, 1987).

Recent advances in radiotelemetry have made it possible to gather detailed knowledge of wolf movements, social organization, and habits. There is now a vast body of detailed information about many aspects of the natural history of wolves.

Common names include gray wolf, timber wolf, and tundra wolf. These are all misleading in the sense that they describe only subsets of *C. lupus*, which is neither uniformly gray, nor restricted to timber or tundra habitat. The simple designator of wolf therefore seems more appropriate.

DISTRIBUTION AND HABITAT

Once widespread in the Northern Hemisphere in both Eurasia and North America, wolves now have a much restricted range. Their original range in North America extended from Central Mexico north to Ellesmere Island (Pimlott, 1975). Currently, the only significant populations in North America are in Canada, Alaska, and northern Minnesota. All populations in the lower 48 states and Mexico are classified as endangered, except in Minnesota where they are classified as threatened. In Eurasia, wolves exist in considerable numbers only east of Europe. Some scattered small populations still remain in Europe and Scandinavia.

Wolves can exist in all temperate habitats and in all types of terrain. They are found in tundra, steppe, forest, plains, and woodland habitats, as well as in mountainous regions up to the boundaries of permanent snow, and along the edges of deserts. They are rare in the taiga (Mech, 1974; Pimlott, 1975; Stroganov, 1969; Novikov, 1962). Mitchell (1977) sighted a wolf at 5,700 m on the north slope of Mount Everest. Dens, often adapted from the previous excavations of other animals, are underground or in similarly protected locations, such as in hollow logs or caves. See Ballard and Dau (1983) for a description of the characteristics of den and rendezvous sites.

PHYSICAL CHARACTERISTICS

Wolves are the largest of the nondomestic Canidae species. There is considerable size variation between individuals from different regions. Females are smaller and lighter than males. Head-plus-body length ranges from 100 to 164 cm, and weight ranges from 18 to 80 kg. The head is relatively large with a broad forehead, muscular neck, and distinctive facial ruff. The chest is deep and broad. The legs are long, enabling a maximum running speed of 55–70 kph. Tails range in length from 30 to 56 cm and are well furred with a dark spot over the caudal gland. Pelage is long and dense. Guard hairs are 60–150 mm in length (Mech, 1974), with a short, dense undercoat. In northern regions of wolf distribution, the undercoat is soft and furry, while in southern areas it is coarser and sparser. Color is highly variable, ranging from white to black, through all intermediate shades of brown, red-brown, sandy yellow, and gray. There is often a dark saddle marking and dark-tipped guard hairs. The underparts are yellow-white, cream, or white. Legs, ears, and muzzle may be paler than the rest of the pelage. The skull is relatively large with a long facial region, massive jaws, and large, almost spherical bullae. The canines and carnassials are large and powerful.

TAXONOMY

Approximately 32 subspecies have been described (Mech, 1974). Since some of these have been described on the basis of only a few specimens, this is probably not a realistic figure (Mech, 1974). Nowak and Paradiso (1983) stated that 24 subspecies are currently recognized, noting that this is also an overgenerous number. A number of subspecies are extinct, including *C. lupus nubilus*, the buffalo wolf of North America. Other subspecies have a tenuous hold (*C. lupus baileyi*, the Mexican wolf), and active intervention is required to rescue them from extinction (McBride, 1980).

Wolves are generally regarded as the recent progenitors of domestic dogs (*C. lupus familiaris*, or *C. familiaris*). Recent biochemical analyses support this view (Wayne and O'Brien, 1987; Simonsen, 1976; Fisher *et al.*, 1976). The karyotype has 2n = 78: NF = 80 (Wurster and Benirschke, 1968). This is the same as in *C. familiaris, C. latrans, C. rufus,* and *C. aureus.*

DIET

Wolves are almost entirely carnivorous and prey primarily on large mammals. Popular tales suggesting more benign proclivities (for example, that they subsist on field mice) are nonsense. Large mammals hunted include moose, caribou, wapiti (North American elk), deer, bison, mountain sheep, and mountain goats (Mech, 1974). A group of wolves is capable of killing a full-grown black bear (Horejsi *et al.*, 1984). Wolves also prey on a number of domestic species. Wolves are able to prey on animals much larger than themselves only because they hunt as a cooperating pack. Adults may consume 9 kg or more of meat at a single feeding, though average daily intake is 2.5–6 kg per wolf per day (Mech, 1974). Other foods include beaver, hares, small rodents, birds, frogs, lizards, fruit, and carrion (Stroganov, 1969; Novikov, 1962). In some regions human garbage is an essential part of the diet (Macdonald *et al.*, 1980).

It is clear that prey population levels are a strongly limiting factor on wolf populations, but the effects of the wolves on the population density of their prey are hotly disputed (van Ballenberghe, 1989; Bergerud and Ballard, 1989; Thompson and Peterson, 1988; Bergerud and Snider, 1988; Theberge, 1990). This is an issue of considerable importance, since wolves prey on game animals and thus compete directly with man for this resource. After prolonged study, it will probably become clear that an array of constantly changing factors limits game levels, and that wolves play a key role. In fact, wolves may constitute a major mortality factor in some predator/prey systems (Mech and Karns, 1977; Pimlott, 1975). The point that large ungulate populations require limiting factors other than food is a crucial one. When food is the primary limiting factor, range deterioration inevitably results. Furthermore, wolves generally prey preferentially on sick, injured, and subadult individuals. In this way they provide continual strong selection pressure, efficiently culling inferior stock. Wolves also provide disease control by their selective predation.

Food caching is well developed, and cache sites are often urine-marked after being dug up. This is also the case in coyotes and red foxes (Harrington, 1981).

ACTIVITY

Most activity is crepuscular or nocturnal, although diurnal activity, particularly during cool weather and in winter, is not uncommon (Afik and Alkon, 1983; Novikov, 1962; Stroganov, 1969; Mech, 1974).

REPRODUCTION

Females are annually monestrous. After a courtship period, mating occurs between January and April, earlier in this time period in southern parts of the range (Fuller, 1989a,b). The mating pair experiences a copulatory tie for up to 30 minutes. Pups are born after a 62–65-day gestation, and litter size ranges from one to 13, with six an average figure. The female usually whelps in a burrow, though other quiet spots are used too. The same den may serve for a number of years in a row. In almost all cases, it is the alpha (dominant) female and alpha male of a pack who mate and produce pups. One instance of polygyny, in which one male mated with two females who both subsequently had pups, has been documented (Mech and Nelson, 1989). The female remains with the pups at the den site for about 2 months, during which time the male and other pack members provision her and the pups. After about 2 months, the pack and pups move to a rendezvous site, or series of rendezvous sites. These are areas of roughly 0.5 acre, which serve as home bases for 1–6 weeks (Carbyn, 1979), until the pups are old enough to travel with the pack. Until then, the pups remain at these rendezvous sites while adults hunt and bring them food (Mech, 1974; Zimen, 1975; Stroganov, 1969; Novikov, 1962). See van Ballenberghe and Mech (1975) for data on pups' physical development and survival. See Harrington *et al.* (1983) for a discussion of alloparental care, pack size, and pups' survival rates.

Most young disperse before the age of 2 years. Some remain with their natal pack, and a few will assume alpha status within their natal pack and breed. Females are sexually mature at 2 years, males at 2–3 years. Sexual maturity is no guarantee of breeding, however, since the alpha pair suppresses breeding in all other pack members. Reproduction in captivity has a high success rate.

In captivity wolves live up to 16 years. Free-ranging individuals in Minnesota may live at least 13 years, with reproduction occurring in 11-year-olds (Mech, 1988). Human hunting is responsible for much mortality in some areas. Disease, parasites, starvation, intraspecific killing, and injuries take their toll as well (Peterson and Page, 1988; Mech, 1974; Novikov, 1962; Roberts, 1977).

SOCIAL ORGANIZATION AND BEHAVIOR

Wolves have the most highly developed social organization of all the Canidae, with the exception of African wild dogs (*Lycaon pictus*) and dholes (*Cuon alpinus*). The essential social unit is the pack, typically composed of a mated pair, their offspring of the year, and a few offspring from previous seasons. Although packs as large as 36 have been reported (Rausch, 1967, cited in Mech, 1974), the usual number is much smaller, typically two to 12. See Zimen (1976) for a study of the regulation of pack size.

Some wolves exist as lone nomads with no fixed home range, either for short periods while dispersing or sometimes for longer. Asian wolves (*Canis lupus chanco*) appear to be solitary, although they sometimes are found in pairs (Mitchell, 1977). Wolves of the Arabian Peninsula hunt singly or in pairs (Harrison, 1968). Thus in some regions, perhaps those where conditions are most demanding, the pack structure is not always present. This contrasts with the social organization of African wild dogs, in which pack structure is without exception obligatory. African wild dogs never exist as lone nomads for extended periods of time.

Wolf packs are characterized by elaborate social interactions. In larger groups, each gender has a separate dominance hierarchy with the alpha male usually dominant over the female (but see Vila *et al.*, 1990). Among pups there are linear dominance hierarchies (Zimen, 1975; Mech, 1974). Pup behavior is distinctly different from adult behavior: Pups, on the whole, are afforded more social leeway than adults (Peterson and Page, 1988). Each pack inhabits and defends an exclusive home range. There is little overlap between the home ranges of neighboring packs, and buffer zones exist between them. Scent marking and howling function to maintain territories, and "direct encounters between packs appear to be of little importance in territory maintenance" (Harrington, 1975, p. 2). High rates of scent marking accompany "territorial adjustments or intrapack expression[s] of dominance" (Peterson and Page, 1988, p. 95). When direct encounters between packs do occur they are marked by chases and fighting, sometimes violent enough to result in death.

Home range sizes are highly variable. They are influenced by food availability and pack size. Areas as large as 13,000 km² (Alaska) and as small as 18 km² (Ontario) have been recorded (Mech, 1974). An adult pair in the Negev occupied a range of 60.3 km² (Afik and Alkon, 1983). Territories in Minnesota are stable in size and location over a period of years (Harrington and Mech, 1979). Lone wolves range through and around these occupied areas. Lone wolves, often dispersing subadults or adults, have a relatively high mortality. Dispersal serves as a population sink, helping to regulate population density. In areas where food supply is adequate but dispersal is limited, wolf densities may reach unsustainable levels, and social violence may result (Peterson and Page, 1988).

Population density is highly variable. The highest recorded is one wolf per 10.8 km² (a nonsustainable density), but it may be as low as one per 520 km² in parts of Canada (Mech, 1974; Nowak and Paradiso, 1983). Population density seems to be determined primarily by food availability. On Isle Royale, Michigan, when food supplies are high, packs are large and dispersal rates are low. As food supply decreases, dispersal rates increase and pack size decreases (Peterson and Page, 1988).

Vocalizations are an important component of social interactions among members of this highly social species. Howls, the most conspicuous of the vocalizations, serve affiliative intrapack and territorial interpack functions. There are a large number of other affiliative and agonistic vocalizations as well (Schassburger, 1978; Harrington, 1975, 1987, 1989; Field, 1978).

Interactions with other canid species in areas of sympatry are described in Macdonald *et al.* (1980) (*C. lupus, V. vulpes*); McBride (1980) (*C. lupus, C. latrans*; Fuller and Keith (1981) (*C. lupus, C. latrans*); and Dekker (1989, 1990) (*C. lupus, C. latrans, V. vulpes*).

Canis mesomelas: Black-Backed Jackal

Common names include black-backed jackal and silver-backed jackal. These are moderately social canids whose natural history is similar to that of the side-striped and golden jackals.

DISTRIBUTION AND HABITAT

There are two disjunct populations of black-backed jackals. The northernmost encompasses southern Ethiopia, southern Sudan, Somalia, Kenya, Uganda, and northern Tanzania. The southern population extends from the Cape of Good Hope northward to Angola, Zimbabwe, and southern Mozambique.

Black-backed jackals typically exist in brushy woodlands (Kingdon, 1977; Moehlman, 1983; Kruuk, 1972a,b; Schaller, 1972). They are also found on open savanna and in open woodlands (Clutton-Brock *et al.*, 1976; van der Merwe, 1953a,b; Dorst and Dandelot, 1969). In the Rift Valley, Kenya, in an area of sympatry with golden and side-striped jackals (*C. aureus* and *C. adustus*), black-backed jackals occurred more often in closed woodland, *C. aureus* in grassland, and *C. adustus* in open woodland (Fuller *et al.*, 1989).

PHYSICAL CHARACTERISTICS

Head-plus-body length is 68–74.5 cm, and height at shoulder is 38–48 cm. Tail length is 26–40 cm. Weight ranges from 6 to 13.5 kg; males are significantly larger than females, averaging 1 kg more (Rowe-Rowe, 1984; Kingdon, 1977). The most distinctive external physical characteristic is the well-defined dark saddle marking, black and silver in color, which extends from the shoulders posteriorly to the root of the tail. The underparts are ginger to cream or white, and the tail, moderately bushy and brownish in color, is black tipped. The head, ears, and lower sides of the body are generally rufous flecked with black and silver hairs (Kingdon, 1977; Dorst and Dandelot, 1969; Clutton-Brock *et al.*, 1976; van der Merwe, 1953b). The skull is smaller than that of golden jackals.

TAXONOMY

According to the analysis by Clutton-Brock *et al.* (1976) of the systematic position of this species, black-backed jackals are more closely related to side-striped jackals (*C. adustus*) than to golden jackals (*C. aureus*). Black-backed jackals appear to be an "exceptionally stable and ancient form," and fossils of the species exist from throughout the Pleistocene (Kingdon, 1977, p. 31). See Wayne *et al.* (1989) for a discussion of divergence among the three sympatric jackal species (*C. aureus, C. adustus, C. mesomelas*). These authors found that the three species are strikingly similar with respect to size-related morphology, much more so than are other sympatric canids, such as South American foxes. This morphological similarity has been retained through at least 2 million years of potential divergence time, in contrast to the relatively rapid morphological divergence of South American canid species. They suggested that the relative diversity of predator and prey species in African ecosystems may in part have served as a damper on the interspecies competition that might otherwise have served as an impetus to morphological divergence.

Genetic analysis (gel electrophoresis of soluble blood proteins) data showed unexpectedly large genetic distance values between *C. mesomelas* and other *Canis* species (*C. lupus, C. latrans, C. familiaris*) (Wayne and O'Brien, 1987). Three subspecies are recognized by Stains (1975).

DIET

Black-backed jackals are opportunistic predators and scavengers. Animals the size of Thomson's gazelles seem to be the upper limit of black-backed jackals' predatory capacity, although they may occasionally kill sick adults of larger species. The relative importance of large prey seems to be seasonally and situationally dependent: Larger prey is captured by groups, while hunting for smaller prey is done by lone individuals (Kingdon, 1977). Van Lawick-Goodall and van Lawick-Goodall (1971) observed that when hunting adult gazelles, black-backed jackals almost always worked in groups of three to seven. This ability to switch from group hunting of large prey to individual foraging is also a characteristic of other canids, such as coyotes. Small

mammals, such as rodents, hedgehogs, mongooses, and young ungulates, are important in the diet. On a game reserve in South Africa,
mammals constituted 75% (by occurrence) of the diet (Rowe-Rowe,
1983). Black-backed jackals consume insects, including beetles,
grasshoppers, termites, crickets, winged ants, and spiders; other prey
includes birds and their eggs, frogs, lizards, snakes, and crabs. They
also eat plant material, such as fruit and groundnuts. Carrion of all
sorts is an important dietary element (Pienaar, 1973; Kingdon, 1977;
Bothma *et al.*, 1984; Dorst and Dandelot, 1969; Meester and Setzer,
1971; van der Merwe, 1953a; Shortridge, 1934; Roberts, 1951). In
coastal regions seabirds, Cape fur seals (in the form of carrion or live
pups), fish, and other marine species constitute almost the entire diet
of coastal-dwelling black-backed jackals (Avery and Avery, 1987; Hiscocks and Perrin, 1987; Stuart and Shaughnessy, 1984).

Black-backed jackals prey on domestic stock and are controlled in
South Africa primarily for this reason (Ferguson *et al.*, 1983; Bothma,
1971a,b; Skead, 1974). Like many other canids, black-backed jackals
cache food (Ewer, 1973). See van der Merwe (1953a) for a detailed list of
diet items. See Avery and Avery (1987) for a discussion of diet in
coastal Namibia, including analysis of jackal middens. These middens,
on vegetated hummocks, are areas where jackals bring food to eat, and
where food remnants such as bones and feathers accumulate. Middens
provide shelter, vantage points, and relatively warm spots in an otherwise harsh environment (Avery and Avery, 1987).

ACTIVITY

In southern Africa, black-backed jackals have a bimodal activity pattern. Both movement and distances traveled are strongly linked to crepuscular hours, with the largest peak in early evening (Ferguson *et al.*,
1988). Their activity patterns thus coincide with those of key prey animals (Ferguson *et al.*, 1988). In a coastal region of Namibia, black-
backed jackals are active throughout the day, with activity peaks at
0900 and 1800 hours (Hiscocks and Perrin, 1988). This flexible, crepus-
cular/diurnal activity pattern is characteristic of jackal populations
unexploited by man. In areas where human disturbances are consistent, black-backed jackals are nocturnal. In protected areas, such as national parks, they are diurnal as well (Kingdon, 1977; Meester and Set-

zer, 1971; Shortridge, 1934; van der Merwe, 1953b). See Ferguson *et al.* (1988) for a discussion of black-backed jackals' activity patterns.

REPRODUCTION

The reproduction of black-backed jackals is similar to that of the other jackal species. Gestation is 60 days (Kingdon, 1977). Van der Merwe (1953a) reported that mature females are capable of producing two litters a year, but this observation awaits supporting evidence. Litters usually number about four, though as many as nine pups have been reported in a single litter (Kingdon, 1977). Meester and Setzer (1971) give a range of six to seven pups per litter; van der Merwe (1953a) a range of five to seven. Pups are born in August or September. Mean litter size in the Natal Drakensberg, South Africa, is 5.6, with pup survival rate of two per litter at the age of 14 weeks (Rowe-Rowe, 1984). Dens are often abandoned aardvark holes or excavated termitaries (Kingdon, 1977). Pups remain at the natal den site until about 12 weeks old, at which time they become more mobile (Ferguson *et al.*, 1983). By the age of 6 months the young may be hunting on their own (Ferguson *et al.*, 1983).

Both parents share in the care of the pups, and they may be assisted in this by the young of previous litters (Ferguson *et al.*, 1983; Moehlman, 1983). Pair bonds are long term: Moehlman (1983) observed pairs that remained together for the duration of 6 years of study. Captives have lived up to 14 years (Kingdon, 1977), although maximum lifespan in the wild is probably closer to 7 years (Rowe-Rowe, 1982). Black-backed jackals are preyed on by leopards (Schaller, 1972).

SOCIAL ORGANIZATION AND BEHAVIOR

Black-backed jackals are moderately social canids (Ferguson, 1978; Ferguson *et al.*, 1983; Moehlman, 1979, 1983). The basic social unit is the mated pair, or the mated pair plus offspring. Moehlman (1979, 1983), in a long-term field study in Tanzania, observed cooperative hunting, food sharing, maintenance of long-term pair bonds and territories, and mutual grooming. Pair bonds last throughout the year. There is little behavioral or physical differentiation between males and females. Social interactions between pair members are consistently amicable, whereas interactions with non–pair members are primarily agonistic (Ferguson, 1978). Ferguson *et al.* (1988) observed that pair members moved in remarkable unison in the Kalahari Gemsbok National Park. In the Natal Drakensberg, South Africa, there appear to be four components of black-backed jackal social organization: territorial mated pairs, their offspring, their (nonreproducing) helpers, and nomads (solitary, nonbreeding, nonterritorial individuals) (Rowe-Rowe, 1982). This population is probably roughly 25% mated pairs, 25% young of the year, and 50% nonbreeding adults (Rowe-Rowe, 1984).

Groups are often observed which seem to consist of mated pairs plus offspring of current and sometimes previous reproductive seasons. Large aggregations of jackals take place at carrion: Schaller (1972) observed a group of 19 at a carcass, and Dorst and Dandelot (1969) stated that groups of up to 30 may gather.

Usually only jackals older than 2 years hold breeding territories. Subadults either remain at their natal den site and act as helpers in the provisioning and guarding of pups, or else disperse (Ferguson *et al.*, 1983). Mated pairs are territorial. They demarcate territories by scent marking, and both members vigorously defend the territory against adult conspecifics (Moehlman, 1979, 1983; Ferguson *et al.*, 1983; Rowe-Rowe, 1982; Kingdon, 1977). Territorial boundaries may be disregarded, however, when a large food source becomes available (Ferguson *et al.*, 1983; Schaller, 1972). In coastal Namibia there is a high degree of home range overlap, which may be related to the abundant, clumped food supply (Hiscocks and Perrin, 1988). In general, clumped, very abundant, nondefensible food supplies are associated with an easing of territoriality in canids. Home range sizes are highly variable (5–180 km²) (Hiscocks and Perrin, 1988). Nomadic individuals are fairly numerous. These may be either dispersing subadults or nonpaired adults of both sexes. Nomads follow herds of migrating herbivores (van Lawick-Goodall and van Lawick-Goodall, 1971). Nonnomadic subadults are often tolerated on the territories of mated pairs (Moehlman, 1983; Rowe-Rowe, 1982; Ferguson *et al.*, 1983). Rowe-Rowe (1982) estimated the mean size of the home range in the Natal Drakensberg, South Africa, at 18.2 km², with a population density of 1 jackal per 2.5–2.9 km².

There is a fairly elaborate repertoire of acoustic communication signals, including growls, whines, cackles, barks, yaps, and howls, as well as graded intervariants of these.

Black-backed jackals are similar to golden and side-striped jackals, but some differences are evident. Moehlman (1983) observed that black-backed jackals have a lower frequency of affiliative behaviors than golden jackals. Kingdon (1977) stated that social behavior seems to be more highly developed in black-backed jackals than in the other jackal species, though exactly what this means is unclear. In the Rift Valley, Kenya, where black-backed, golden, and side-striped jackals are sympatric, habitat and temporal activity segregation appear to limit intraspecific competition (Fuller *et al.*, 1989). See Bothma *et al.* (1984) for information on the ecology of the Cape fox (*Vulpes chama*), bat-eared fox (*Otocyon megalotis*), and black-backed jackal in an area of sympatry in the Namib Desert. Pronounced similarities between black-backed jackals and coyotes (*C. latrans*) are obvious in natural history, social organization, and relationship to man.

Canis rufus: Red Wolf

It is by no means clear that the red wolf is a distinct species. There has been considerable debate on the issue for at least 20 years. The general opinion at this time favors specific status, with a few persuasive dissenting opinions (see the taxonomy section below). Red wolves are listed as endangered by the IUCN, and are on Appendix I of CITES (Ginsberg and Macdonald, 1990). They are extinct in the wild except for several small reintroduced populations. Their decline is thought to be due to a complex of factors, including habitat destruction, aggressive long-term control programs, hybridization with coyotes, and high mortality from susceptibility to parasites (Parker, 1988; Paradiso and Nowak, 1971, 1972; Carley, 1979; Ferrell *et al.*, 1980).

In 1967 red wolves were listed as endangered by the U.S. government. During the early 1970s, research efforts were focused on taxonomy, censusing, and location. In 1973 the Endangered Species Act was passed, and red wolves were selected for priority treatment (Carley, 1979). In the same year, a captive breeding program was established. Its goals were to identify wild-caught red wolves and to establish a breeding population for distribution to other facilities and eventual reestablishment in the wild. During the 1970s about 40 wild-caught canids, supposed to represent *C. rufus*, were brought into captivity. Of these, 17 were eventually certified as pure *C. rufus* (Parker, 1988). By late 1975 it was clear that species preservation in the wild was impossible (Carley, 1979). In the mid to late 1970s, efforts centered on the capture of the few remaining wild red wolves and their integration into the captive breeding program. The first captive litters were born in 1977. In 1980 the U.S. Fish and Wildlife Service declared red wolves extinct in the wild, and location and capture efforts were terminated. The entire gene pool on which the resurrection of the species depended was embodied in between 17 and 30 individuals. By any estimation, the genetic bottleneck was excruciatingly severe. Nonetheless, the captive population has increased steadily; as of fall 1990, it stood at 131. Captives are held at 22 separate facilities.

In 1988, eight captive-bred individuals were released in the Alligator River National Wildlife Refuge, North Carolina. Two pairs reproduced the following spring. As of fall 1990, there were 9 free-ranging and 21 captive red wolves on the refuge, with some of the captives due for release in the autumn (Parker, 1990). The project's goal is to have 25–35 individuals on the refuge in 1992. Despite considerable, often human-caused mortality, the project is considered to be going well. Plans for reintroduction to the 500,000-acre Great Smoky Mountains National Park are moving forward. The plan is to release several pairs of adults in the park in 1991. A potential stumbling block is the presence of a low density coyote population (about 90% of the former range of the red wolf is now occupied by coyotes) (Parker, 1990).

DISTRIBUTION AND HABITAT

Research reports from the 1960s and 1970s document a rapidly diminishing range and decreasing population. Incursion of coyotes into the range of the red wolf and concomitant hybridization sealed the fate of red wolves. By May 1977 only a few individuals were present in isolated areas (McCarley and Carley, 1976). By the time anyone paid much attention to the issue of red wolf distribution, the species was in such an endangered state that determining their range became an exercise in detective work and historical reconstruction. The best available evidence indicates that red wolves were once widely distributed through the southeastern United States from southern Florida to central Texas, and possibly as far north as the Carolinas and Kentucky (Paradiso and Nowak, 1971, 1972; Carbyn, 1987). By 1940 red wolves were absent from Missouri and Oklahoma (where they were replaced by coyotes) and from Arkansas and most of Texas and Louisiana (where they were replaced by red wolf x coyote hybrids) (Paradiso and Nowak, 1971). By 1970, the red wolf range had shrunk to a few thousand square kilometers in southern Texas and southwest Louisiana. This range implosion was simultaneous with coyote range expansion and red wolf x coyote hybridization (Paradiso and Nowak, 1971). See Paradiso and Nowak (1972) and Riley and McBride (1972) for distribution maps.

On their historic range, habitat included forests, coastal prairies, and marshland. Apparently red wolves preferred a "warm, moist, densely vegetated habitat" (Paradiso and Nowak, 1972 p. 3). In their final range, heavy cover seemed to be a requirement (Carley, 1979). At the last, virtually all red wolf habitat was on privately owned land: Grazing, agriculture, and petrochemical industrialization were the primary uses of this land (Riley and McBride, 1972; Carley, 1979). The climate in this region was subtropical with high relative humidity—habitat "markedly different from the habitat found over the majority of its historic range" (Riley and McBride, 1972 p. 1).

PHYSICAL CHARACTERISTICS

Red wolves are intermediate in size and physical characteristics compared to coyotes and gray wolves. There may be some size overlap at both ends of the spectrum. Total length is 135–165 cm; tail length is 25–35 cm. Weight is 16–41 kg (Paradiso and Nowak, 1972; Ginsberg and Macdonald, 1990). Early trapping records indicated that the largest *C. rufus gregoryi* specimens weighed up to 36 kg (Carley, 1979). The name *red wolf* is a misnomer since coat colors are highly variable, including shades of cinnamon, tawny gray, rufous gray, rufous, and, rarely, black. Tawny with gray and black highlights is a common pelage coloration. On the whole, pelage tends to be more rufous, and more sparse, than is generally the case for coyotes or gray wolves (Paradiso and Nowak, 1972). Underparts are lighter in color, whitish to buff. There is a conspicuous black caudal gland marking. The legs are proportionately long and slender (Parker, 1988). There is a conspicuous narrow black marking on the anterior surface of the foreleg, as in coyotes and wolves (Carley, 1979). Interestingly, Carley (1979, p. 15) remarked that "after a period of time, red wolves shipped to the captive breeding program in Tacoma, Washington, change color, put on a heavier coat, and become somewhat grayer with the white on the chest, legs, feet, and muzzle becoming more pronounced." Facial coloration patterns are similar to those of gray wolves. The muzzle is light in color, and there are often light-colored areas around the eyes. There may be a tan-colored spot over each eye (Riley and McBride, 1972).

Red wolves and coyotes are physically very similar. Riley and McBride (1972, p. 3) asserted that, on the basis of external appearance alone, the two can be "readily distinguished in the field." Carley (1979) stated that the two species are readily distinguishable, and that difficulty arises only when dealing with hybrid forms. *C. rufus* is always larger cranially and overall than the same sex of *C. latrans* (Paradiso and Nowak, 1972). Riley and McBride described a number of differing traits, such as width of muzzle and nose pad, head breadth, ear proportion, and angle subtended by the ears to the head. In addition, red wolf tracks are larger and stride length is longer than in coyotes. Scat diameter is larger as well.

When compared with those of gray wolves, red wolf skulls are slenderer with a larger rostrum and narrower, longer canines. The braincase is smaller and more ossified, and the sagittal crest more prominent (Paradiso and Nowak, 1972; Goertz et al., 1975). See Lawrence and Bossert (1967, 1975) for details of skull differences in *C. rufus*, *C. lupus*, *C. latrans*, and *C. familiaris*. The dental formula in *C. rufus* is the same as for all members of the genus *Canis*: incisors 3/3, canines 1/1, premolars 4/4, molars 2/3 = 42.

TAXONOMY

The taxonomic status of red wolves has been marked by confusion and disagreement. There are four views: (1) red wolves are a distinct species; (2) red wolves are descended from coyote x gray wolf hybrids, and should not be treated as a separate taxon; (3) red wolves are a subspecies of gray wolf; (4) red wolves hybridized with coyotes to such an extent during the recent past that they should not be treated as a separate taxon. The majority opinion at the present time recognizes the species as a valid taxon (Wozencraft, 1989; Ginsberg and Macdonald, 1990; Carbyn, 1987; Parker, 1988; Paradiso, 1968; Paradiso and Nowak, 1972; Shaw, 1975). Dissenting opinions are represented by

Ewer (1973), Lawrence and Bossert (1967, 1975), and Clutton-Brock *et al.* (1976). Lawrence and Bossert (1967, 1975) presented a compelling case for red wolves being only subspecifically distinct from gray wolves, as well as an abbreviated history of the taxonomic issues involved.

The origin of *C. rufus* is unclear. There is no fossil record (Paradiso and Nowak, 1972). The range of *C. lupus* apparently did not extend into the southeastern United States in the recent past (Kennedy, 1987). Therefore, it seems unlikely that red wolves are the descendants of gray wolf x coyote hybrids (but see Mech, 1970). Discovery of a skull and postcranial remains in an Alabama cave appear to support the longevity of red wolf status in the Southeast (Paradiso and Nowak, 1972).

Indisputably, extensive hybridization between red wolves and coyotes has occurred. There is evidence of hybridization in central Texas as early as 1915–18 (Paradiso and Nowak, 1971), and extensive hybridization was probably occurring by the mid-1960s (McCarley and Carley, 1976). This hybridization is attributed to the malign human influences of habitat disruption combined with species persecution. The determination that coyotes and red wolves are separate species is based on the observation that the physical differences between red wolf specimens and coyotes are greater than the differences among coyote subspecies (Carley, 1979). Nonetheless, when putative pure red wolves were first brought into captivity, a battery of measurements was needed to determine the taxonomic status of each individual. In a sense, the question of whether *C. rufus* was a distinct species is moot: After forcing fewer than 30 of these individuals through a genetic bottleneck, humans are in the process of constructing a new form of *Canis*.

Red wolves were known as *C. niger* Bartram, 1791 until 1967, when that designation was invalidated in favor of *C. rufus* Audubon and Bachman, 1851. There were three subspecies: *C. r. floridanus* (eastern), *C. r. rufus* (western), and *C. r. gregoryi* (central). Both *C. r. floridanus* and *C. r. rufus* were extinct by the mid-1900s (Parker, 1988; McCarley, 1962). Despite some preliminary work by Ferrell *et al.* (1980), biochemical/genetic markers particular to red wolves have yet to be found. Red wolf karyotype (2n = 78; NF = 80) is the same as in coyotes and wolves (Wurster and Benirschke, 1968).

DIET

Red wolves, like coyotes, are opportunists. Small animals, such as rabbits, raccoons, and nutria, are their primary prey. They consume fish, insects, carrion, and plant material as well (Paradiso and Nowak, 1972; Carley, 1979; Riley and McBride, 1972; Shaw, 1975). Only occasionally do they prey upon ungulates. There are a few historical records of three or more red wolves attacking cattle, deer, or razorback hogs (Paradiso and Nowak, 1972). Predation on new calves does occur occasionally (Carley, 1979). The extermination efforts of the early and mid-1900s were justified by red wolf predation on smaller domestic animals, such as free-ranging hogs (Riley and McBride, 1972).

ACTIVITY

Activity is bimodal and nocturnal. There is a primary peak at 2000–0000 hours, followed by inactivity from 0100–0300, when activity resumes until dawn (Carley, 1979; Shaw, 1975). During daylight hours, red wolves usually rest in cover (Paradiso and Nowak, 1972). In winter, increased diurnal activity occurs. See Roper and Ryon (1977) for an investigation of activity patterns in captive red wolf x coyote hybrids.

REPRODUCTION

Females are annually monestrous, with most breeding in their second year. Mating occurs from late December to early March, and pups are born in March, April, and May. Gestation is 60–62 days (Riley and McBride, 1972; Carley, 1979; Paradiso and Nowak, 1972). Litters range from two to eight, with an average of three or four (Riley and McBride, 1972; Parker, 1988). In captivity, wild-caught captives mated successfully after being placed together for 2–3 months (Carley, 1979). Both parents in mated pairs care for and play with their pups (Carley, 1979). On the whole, reproduction in captivity has been successful, although artificially inseminated females have so far failed to bear pups.

Human-caused mortality has been responsible in large part for the decline of red wolves. Apparently red wolves lacked the caution of coyotes and were therefore more readily trapped and poisoned (Paradiso and Nowak, 1972). Road kills are also a significant cause of mortality. Parasites exact a heavy toll. Of 27 wild-caught wolves tested, all 27 had heartworm (Riley and McBride, 1972). Intestinal parasites, distemper, and mange are also widespread (Riley and McBride, 1972; Paradiso and Nowak, 1972). The high parasite burden carried by all red wolves may indicate that they were occupying marginally suitable habitat. The majority of animals captured during the intensive capture efforts of the 1970s were less than 4 years old (Carley, 1979), indicating a very high mortality rate for older individuals. Paradiso and Nowak (1972) noted that there appeared to be very low levels of pup survivorship on the Texas gulf coast in the late 1960s, with most pups dying before 6 months of age. Potential lifespan, if comparable to that of free-ranging coyotes, should have been as much as 12 years.

SOCIAL ORGANIZATION AND BEHAVIOR

The social organization of red wolves before their slide toward extinction will never be known. It is probable that whatever form their social organization had, it could not withstand the combined pressures of habitat destruction, intense persecution, introgression with coyotes, and relegation to suboptimal habitat. No formal studies of social organization were done before the species was on the ropes, so a description of these wolves' social organization is, by necessity, an exercise in historical reconstruction. It seems that red wolves had a social organization more similar to that of coyotes than of gray wolves. Red wolves traveled in groups of up to seven. Groups were larger in the fall, when offspring and parent groups traveled together. Lone wolves were also often spotted (Carley, 1979; Shaw, 1975). Paradiso and Nowak (1982) reported that aggregations of 5–11 wolves sometimes gathered briefly, but after greeting one another, they broke up again into family groups. Riley and McBride (1972, p. 9) stated that red wolves maintained a group structure year-round. They observed groups of three or more traveling together. As with coyotes, young of the previous year may have stayed near dens, although they apparently did not provide alloparental care (Riley and McBride, 1972). Mated pairs, sometimes accompanied by an "extra" male, traveled together (Paradiso and Nowak, 1982, p. 3). Scratch and scent marks occurred along habitual travel routes (Paradiso and Nowak, 1982). Observations of captives show that pair mates display a range of canid affiliative behaviors

typical of *Canis* species. They often greet and play with one another
(Carley, 1979). See Ryon (1979) for a description of social organization
in captive red wolf x coyote hybrids.

Home ranges of red wolves are estimated to have been 40–80 km^2
(Riley and McBride, 1972). Males' home ranges were at the larger end
of this range, those of females at the smaller (Carley, 1979).

See McCarley (1978) and McCarley and Carley (1976) for descriptions of red wolf vocalizations. Like so much else about red wolves, apparently the characteristics of their vocalizations fall between those of coyotes and gray wolves, slightly favoring the coyote end of the spectrum (Paradiso and Nowak, 1972).

Canis simensis: Ethiopian Jackal

This rare canid is variously known as the Ethiopian jackal, Simien jackal (Clutton-Brock *et al.*, 1976), Abyssinian jackal (Wendt, 1975b), Abyssinian wolf (Dorst and Dandelot, 1969), and Simien or Semien fox (Tyler, 1975; Brown, 1964; Bolton, 1973). It is not closely linked taxonomically to the foxes (genus *Vulpes*) and therefore ought not be referred to as a fox. It is more similar behaviorally and physically to jackals than wolves and occurs exclusively in Ethiopia, so the name Ethiopian jackal is most appropriate.

DISTRIBUTION AND HABITAT

This is a montane species, which inhabits high plateaus in the Simien, Balé, and Arussi mountains in Ethiopia. A small population has also been found in the mountains of northwest Shou (the central province of Ethiopia) (Tyler, 1975). The species is not found outside these areas (Morris and Malcolm, 1977; Clutton-Brock *et al.*, 1976; Tyler, 1975).

Unlike most canids, Ethiopian jackals have a range restricted to high altitudes. They are found above 2,900–3,000 m (Morris and Malcolm, 1977; Clutton-Brock *et al.*, 1976). Yalden *et al.* (1980) stated that there are no recent records of the species at altitudes much below 3,000 m. In the Simien Mountains, Brown (1964) found sign and observed animals from 3,125 m to 4,062 m, the limit of vegetation. These jackals are also found in montane grasslands and moorland (Yalden *et al.*,

1980; Morris and Malcolm, 1977; Brown, 1964). See Emmrich (1985) for photographs and brief descriptions of habitat.

The numbers of the Ethiopian jackal have declined drastically in the Simien and Arussi regions during past decades (Yalden *et al.*, 1980; Bolton, 1973). The Balé Mountains population probably numbers about 350–475 individuals (Morris and Malcolm, 1977); perhaps fewer than 20 individuals remain in the Simien Mountains (Tyler, 1975). Thus there is no doubt that the species is both less numerous and less widely distributed now than in the past. The primary causes of this decline are human hunting and progressive habitat destruction (Yalden *et al.*, 1980; Emmrich, 1985). Ethiopian jackals are classified as endangered (Macdonald, 1984) and are protected by law in Ethiopia (Morris and Malcolm, 1977). Nonetheless, they are moving closer to extinction, and a road now traverses the once isolated area of their highest population density (Morris and Malcolm, 1977; Tyler, 1975).

PHYSICAL CHARACTERISTICS

Head-plus-body length averages 99 cm, with a shoulder height of 60 cm, and a tail length of 25 cm (Mivart, 1890; Emmrich, 1985). Ethiopian jackals are slightly larger, longer legged, and rangier than European red foxes, and they have very long and slender muzzles (Tyler, 1975; Mivart, 1890). Pelage color is rufous, with pale ginger underfur and whitish underparts (Clutton-Brock *et al.*, 1976). There is a conspicuous white marking across the chest, and the chin and throat are also white (Emmrich, 1985; Dorst and Dandelot, 1969). The tail is thickly furred, with a conspicuous black plume. The skull is jackallike with an elongated facial region. The teeth in general, and particularly the upper carnassials, are small (Clutton-Brock *et al.*, 1976). Ethiopian jackals have been described as standing out conspicuously against the gray-green vegetation of their habitat (Tyler, 1975), and they certainly do not seem at all camouflaged in photographs (Morris and Malcolm, 1977; Tyler, 1975; Bueler, 1973).

TAXONOMY

Van den Brink (1973) suggested that the nearest relatives of the Ethiopian jackal are in South America, a contention that is weakly supported by data from the findings of Clutton-Brock *et al.* (1976, p. 150), which show a "seemingly close similarity with the genus *Dusicyon.*" Clutton-Brock *et al.* rightly point out that a paucity of data on both postcranial skeleton and behavior may contribute to this apparent similarity. Van den Brink (1973) suggested that *C. simensis* already existed as a distinct form when the genus *Canis* was still undergoing speciation at the beginning of the Tertiary, and that it is even more primitive than the golden jackal (*C. aureus*). Clearly, more information is needed to clarify the taxonomic position of this species.

DIET

The diet of Ethiopian jackals consists primarily of diurnal rodents, which are "astoundingly numerous" in the Balé Mountains (Malcolm and Morris, 1977, p. 154). The giant mole rat forms the bulk of the prey (Morris and Malcolm, 1977; Tyler, 1975; Brown, 1964). Hares are also eaten (Morris and Malcolm, 1977). Although the local people insist that this species kills and eats domestic stock, no substantive supporting evidence of this has been reported (Morris and Malcolm, 1977; Yalden *et al.*, 1980; Brown, 1964). Nonetheless, human retaliation for alleged sheep killing is responsible in part for population decline.

ACTIVITY

In the Balé Mountains, Ethiopian jackals are primarily diurnal, although there is evidence of some nocturnal activity (Morris and Malcolm 1977; Dorst and Dandelot, 1969; Wendt, 1975b). In the Simien Mountains, where the population verges on extinction, the few remaining individuals are almost entirely nocturnal (Morris and Malcolm, 1977). This shift to nocturnal activity patterns when human activity impinges beyond a certain point occurs in other canid species as well (i.e., coyotes, golden jackals, side-striped jackals, wolves).

REPRODUCTION

Pair formation happens in January, and pups are born no later than May or June (Morris and Malcolm, 1977). Morris and Malcolm (1977), who studied Ethiopian jackals in the Balé Mountains, never saw any pups and were unable to locate any dens.

SOCIAL ORGANIZATION AND BEHAVIOR

Ethiopian jackals are solitary to moderately social canids. They have been observed singly, in pairs, and in small groups probably composed of parents and their offspring. Brown (1964) reported that all the animals he saw were either single or in pairs, with the exception of one group of four: a male, a female, and two nearly grown pups. Tyler (1975) reported seeing a group of three, probably a family, as well as a pair, and three lone animals. Morris and Malcolm (1977) saw groups of up to seven, but 65% of their sightings (180 total) were of lone animals. Fiennes and Fiennes (1969) stated that Ethiopian jackals hunt in packs, but no other authors have reported any evidence of cooperative or group hunting, and there would not seem to be much advantage to this behavior since the primary food source, small rodents, is neither large nor defensible. Morris and Malcolm (1977) stated that all hunting activity they observed was by lone animals.

Although all Ethiopian jackals seem to hunt alone, hunting ranges of individuals overlap considerably. Morris and Malcolm (1977) observed up to four animals hunting simultaneously within an area of 2 km^2. No direct evidence for territoriality was observed by these authors, but marking and vocalizing behaviors suggested the presence of territories, and twice they saw lone jackals move away from an area rapidly "to avoid a group."

A wide range of affiliative behaviors occur. These include greeting ceremonies near dawn and dusk, when most sightings of groups are made, as well as "elaborate and friendly social interactions" (Morris and Malcolm, 1977, p. 157). Play, mutual grooming, group vocalizations, and groups of up to four individuals resting in contact have also been observed. Groups also travel together (Morris and Malcolm, 1977).

Tyler (1975, p. 564) characterized the Ethiopian jackals she observed as "extraordinarily tame" and "very inquisitive." They approached her vehicle to within 3 m. Morris and Malcolm (1977) also noted that the animals they observed were quite tolerant of human presence. This tameness, while rendering the jackals good subjects for field study, also makes their extinction all the more likely, since they lack the wariness necessary for survival.

Crab-Eating Fox (*Cerdocyon thous*)
Credit: Kenneth W. Fink/National Audubon Society
Collection/Photo Researchers, Inc.

CHAPTER 4
Genus *Cerdocyon*

Cerdocyon thous: Crab-Eating Fox

The generic name of the crab-eating fox derives from the Greek *kerdo*, meaning "fox", and *cyon*, meaning "dog." The specific name *thous* is derived from the Greek for jackal (Berta, 1982). This South American species does in fact embody a number of characteristics of jackals, dogs, and foxes in its social structure, life history, behavior, and physical characteristics. Crab-eating foxes are fairly common and not particularly elusive.

DISTRIBUTION AND HABITAT

Crab-eating foxes occur across a large region of middle and northern South America. They are found throughout the Brazilian subregion, with the exception of the Amazon Basin Lowlands, and from Colombia and Venezuela southward through Brazil into northern Argentina and Uruguay (Berta, 1982; Langguth, 1975c).

Preferred habitats range from dense tropical forests and woodlands to open grassland. These seem to be primarily forest-dwelling animals (Langguth, 1975c), but they are also found at forest edges, on the llanos [described by Brady (1979) as open palm-shrub habitat], and in a variety of open or gallery forest habitats as well.

PHYSICAL CHARACTERISTICS

In comparison with the family Canidae as a whole, members of this species are mid-sized. Head-plus-body length is 60–70 cm, tail length is 30 cm, and weight is 6–8 kg (Brady, 1979; Langguth, 1975b). The limbs are shorter and more robust than is general for the Canidae, a characteristic that probably facilitates movement in the dense tropical forests that members of this species may inhabit. The ears are relatively short. The long tail (about one half the total head-plus-body length) is a feature shared by the vulpine members of the family Canidae.

Pelage color is variable, generally grizzled brown to gray, often with yellow or white hairs interspersed, particularly along the dorsal midline (Berta, 1982; Langguth, 1975b; Stains, 1975). Face, ears, and legs are tawny to rufous. The underparts are lighter in color, brownish-white to whitish. The tail, long and bushy, is tipped in black or dark brown (Brady, 1979; Berta, 1982; Clutton-Brock *et al.*, 1976; Langguth, 1975b).

TAXONOMY

The taxonomic position of this species is unclear. Evidence from both recent and fossil forms supports generic status in the genus *Cerdocyon* Hamilton Smith, 1839 (Berta, 1982, 1987). Clutton-Brock *et al.* (1976, p. 175), however, while recognizing that "the species lies somewhat apart from the main *Dusicyon* group for some characters, for example the somewhat enlarged frontal sinuses and dark pelage," placed crab-eating foxes in the genus *Dusicyon*. Wozencraft (1989) and Macdonald (1984) also placed them in *Dusicyon*. Van Gelder (1978) placed *Cerdocyon* in a monotypic genus of *Canis*. There are seven subspecies (Macdonald, 1984). As in a number of other South American canids, the karyotype has 2n = 74, and NF = 110. This is an anomalously high NF, indeed the highest recorded in all the Carnivora (Brum-Zorilla and Langguth, 1980). Wayne and O'Brien (1987) found *Pseudalopex vetulus* (which they refer to as *Dusicyon vetulus*) and *Cerdocyon thous* to be closely associated with each other (as sister taxa), and rather more distantly related to the other South American canids. Berta's (1987) cladistic analysis placed *Cerdocyon*, *Atelocynus*, *Speothos*, and *Nyctereutes* within the *Cerdocyon* clade.

DIET

Crab-eating foxes are omnivorous opportunists. Small rodents and domestic fowl constitute the upper size limit of their predatory capability (Biben, 1982a,b; Langguth, 1975b,c; Coimbra Filho, 1966; Berta, 1982; Montgomery and Lubin, 1978; Brady, 1979). Vertebrate prey includes small mammals, particularly rodents, and lizards, frogs, and birds. Depredations on domestic fowl are reported by ranchers. Crab-eating foxes consume invertebrates, including grasshoppers and other insects, large snails, and the eponymous land crabs. They also eat vegetable material, such as maize, figs, small berries, bananas, mangos, and fruit of llanos palms. They consume carrion, including that of domestic stock, as well as roadkills and a variety of scavenged material, including human refuse. Eggs of iguanas and turtles may be an important and concentrated food source (Brady, 1979; Montgomery and Lubin, 1978; Biben, 1982a,b; Langguth, 1975b,c; Coimbra Filho, 1966).

Based on observations of free-ranging foxes in the Venezuelan llanos, Brady (1979) reported seasonal variability in diet, with a heavy reliance on insects and fruit during the wet season, when the foxes hunted on higher ground. During the dry season, they foraged principally on the lowlands where crabs and other vertebrates were more abundant and assumed a greater importance in the diet.

ACTIVITY

Crab-eating foxes are chiefly nocturnal (Berta, 1982; Langguth, 1975b; Allen, 1923). Brady (1979) observed free-ranging individuals in Venezuela foraging from 1800 to 2400 hours, although these individuals also rested intermittently during this time. Brady also observed brief periods of daytime activity.

REPRODUCTION

Little information is available on crab-eating foxes' reproductive behaviors in the wild, but Brady (1978) gave a detailed report on reproduction in captives. Crab-eating foxes form stable, monogamous pair bonds, and both parents share in pup care and den guarding (Brady, 1978). Captive females appear either to have two estrous periods a year or to be reproductively aseasonal, with estrus occurring at any time during the year (Porton *et al.*, 1987). Semiannual or aseasonal estrous periods are a rare, although not unique, reproductive pattern in the Canidae. They may also occur in bush dogs (*Speothos venaticus*), African wild dogs (*Lycaon pictus*), and golden jackals (*Canis aureus*) (Porton *et al.*, 1987).

Captive females may breed in their first year of life. Gestation is roughly 56 days, though it varies considerably (Brady, 1978). In the wild in Venezuela, the breeding and whelping seasons are not well defined (Montgomery and Lubin, 1978). Litters usually number from three to six (Langguth, 1975b). Females may have litters every 8 months (Brady, 1978). As is common for many other canid species, crab-eating foxes occupy the abandoned burrows of other animals, and they have not been reported to excavate their own (Brady, 1979). The young disperse at about 5–8 months of age (Brady, 1979).

SOCIAL ORGANIZATION AND BEHAVIOR

The fundamental social unit is the monogamously mated pair. Crab-eating foxes are moderately social, tending toward the less social end of the spectrum of moderate sociality in the family Canidae. The mean group size observed in Venezuela (202 group sightings) was 1.8 (Montgomery and Lubin, 1978). Roughly one quarter of these sightings were of lone animals; half were of pairs. The largest group seen was composed of five animals and was probably, like the other large groups, a family unit.

Crab-eating fox pups in captivity rested within 1m of each other or their parents for the first months of life (Biben, 1983). In comparison, less social canid species, such as maned wolves (Chrysocyon brachyurus), begin to rest several meters apart by the end of their second month (Biben, 1983). The males of mated pairs assist in parental care and den guarding, and they may provision pups and pregnant or lactating females as well (Kleiman and Brady, 1978). Mutual grooming, sniffling, and licking the mate's head all take place (Brady, 1979; Kleiman and Brady, 1978). While foraging, pair members travel within 30m of one another, and neither pair member takes the lead consistently (Montgomery and Lubin, 1978; Brady, 1979). In this way pair mates hunt together, but their hunting is only very rarely cooperative. Pair mates do interact frequently as they forage together, by urine-marking the same spot sequentially, with either pair member initiating these markings (Brady, 1979; Montgomery and Lubin, 1978).

The frequency of affiliative intrapair interactions is low. Pair mates observed in the field generally interacted with each other only at the end of a rest period or after a period of separation (Brady, 1979). No group affiliative ceremonies, such as group vocalizations or group greetings, have been reported in either wild or captive crab-eating foxes. No instances of alloparental care have been recorded either.

Territorial boundaries are not stringently defended or observed, at least not in the Venezuelan llanos where the only field studies of this species have been conducted. Home ranges of a number of pairs overlapped (Brady, 1979; Montgomery and Lubin, 1978). Several adult pairs may forage through a single area within 15 minutes of each other (Montgomery and Lubin, 1978). Brady (1979, p. 165) observed males from neighboring home ranges engaging in brief agonistic interactions, including wrestling and chasing, but remarked that "intergroup encounters appeared to be more aggressive during the dry season." Perhaps this seasonality is related to the seasonal changes in resource availability that Brady observed. Brady (1981) has compiled a vocal repertoire for this species.

Maned Wolf *(Chrysocyon brachyurus)*
Credit: Francois Gohier/National Audubon Society Collection/Photo
Researchers, Inc.

CHAPTER 5
Genus *Chrysocyon*

Chrysocyon brachyurus: Maned Wolf

Maned wolves, solitary and elusive, are among the least social members of the family Canidae. The size of world populations is unknown (Brady and Ditton, 1979; Meritt, 1973). The most recent estimate is da Silveira's (1968) figure of 1,500–2,200 individuals in 650,000 km² in Brazil. Approximately 100 individuals are exhibited in zoos (Dietz, 1985). Dietz (1981, 1984), who has recently completed the most thorough field studies of the social organization and ecology of the maned wolf, suggested that the species is not in immediate danger of extinction over most of its range. He noted, however, that this condition may change radically, depending on the extent and rapidity of impending habitat destruction. Maned wolves are classified as vulnerable by the IUCN, and are listed as endangered by agencies of the Brazilian government (Dietz, 1984; Thornback and Jenkins, 1982). Study and censusing of maned wolves are complicated by their extremely elusive nature. In two years of field work, Dietz (1984) saw maned wolves on only five occasions without the aid of radiotelemetry, and three of these encounters were with individuals preying on domestic chickens in Dietz's yard.

DISTRIBUTION AND HABITAT

Maned wolves are found in eastern central South America from northern Argentina north into eastern Bolivia and Paraguay, and across into eastern, southern, and central Brazil (Langguth, 1975c; Hershkovitz, 1972, cited in Clutton-Brock *et al.*, 1976; Stains, 1975). See Dietz (1985) for a distribution map. See Thornback and Jenkins (1982) for detailed, region-by-region distribution. The typical habitat includes grassland (grassy savannas and pampas), forest edges, dry shrub forests, swampy areas, and terrain bordering rivers (Thornback and Jenkins, 1982; Langguth, 1975b,c; Stains, 1975; Clutton-Brock *et al.*, 1976).

PHYSICAL CHARACTERISTICS

Maned wolves are the largest of the South American canids. Head-plus-body length is 100–125 cm, height at shoulder is 75–90 cm, and tail length is a proportionately short 30–45 cm. Weight ranges from 20 to 23 kg (Clutton-Brock *et al.*, 1976; Dietz, 1981, 1984, 1985; Langguth, 1975b; Stains, 1975). Their physical appearance deviates significantly from the typical appearance of the nonspecialized canids. Maned wolves have extremely long legs, which give them a vertically elongated aspect unique among the canids. The pelage is rusty red, and there is a long erectile mane on the shoulders and the back of the neck. Adults do not have a layer of underfur. The throat, insides of ears, and tail tip are white, and the muzzle and lower legs are dark to black (Dietz, 1984, 1985; Langguth, 1975b; Clutton-Brock *et al.*, 1976; Stains, 1975). Pelage color varies very little between individuals: Of nine maned wolves studied in Brazil, Dietz (1984) detected no discernible (to humans) differences in this characteristic. Kleiman (1972) noted that there was no detectable caudal gland on the individuals she examined. The skull is large and elongated, the auditory bullae are very small, and the teeth are "simple, widely spaced and 'fox-like'" (Clutton-Brock *et al.*, 1976, p. 177). The canines are long and slender, the upper incisors weak (Dietz, 1985).

TAXONOMY

Chrysocyon brachyurus is the only species in the genus *Chrysocyon* Hamilton Smith, 1839. The maned wolf clearly stands alone as a taxonomic entity. The analysis of the physical characteristics of maned wolves by Clutton-Brock *et al.* (1976) shows closest similarity with the Ethiopian jackal (*Canis simensis*), and these authors suggested that the maned wolf's true taxonomic affinities lie in a low level alliance with the genus *Dusicyon*

Although somewhat foxlike in appearance and habits, maned wolves are not closely allied with the foxes, and they have the round eye pupils characteristic of nonvulpine members of the family Canidae. Characteristics that differentiate them from *Canis* are their very elongated limbs, a short straight cecum, and carnassials that are relatively small compared to the cheekteeth (Berta, 1987). Karyotype is 2n = 76: NF = 78 (Brum-Zorilla and Langguth, 1980). No subspecies are recognized (Clutton-Brock *et al.*, 1976; Dietz, 1984; van Gelder, 1978; Stains, 1975). Based on the results of electrophoretic analyses, Wayne and O'Brien (1987) found that *Chrysocyon* is not closely related to any of the other canids studied (all but *Cuon* and *Atelocynus*). *Chrysocyon* "appears to represent the sole terminal species of a 6-million-year-old lineage. No fossil or living intermediates exist to connect this morphologically aberrant species with ancestral fossil forms" (Wayne and O'Brien, 1987, p. 348). Maned wolves thus appear to be the sole survivors of the late Pleistocene extinction of large canids in South America.

DIET

Maned wolves are omnivorous. They hunt and scavenge alone, relying on relatively small and dispersed food items. In the Serra da Canastra National Park, Brazil, the single most important food item is the fruit of *Solanum lycocarpum*, which resembles large tomatoes. Based on scat analysis, this fruit is the predominant diet component in both volume and frequency of occurrence. Small mammals (particularly rodents), vegetable items such as various fruits, and grass also figure prominently in the diet. Maned wolves eat birds and insects (Dietz, 1981, 1984; Bartmann and Bartmann, 1986), as well as reptiles, gastropods and other terrestrial mollusks, bird eggs, bananas, guavas, bulbs, and roots (da Silveira, 1968). Diet may vary with the seasons (Dietz, 1984). Maned wolves do not seem to prey to any great extent on domestic stock, although they may rarely take newborn lambs or pigs (Dietz, 1984). They do, however, frequently kill domestic poultry, a habit indirectly responsible for wolf mortality, since it leads to human retribution (Dietz, 1984).

ACTIVITY

Free-ranging maned wolves are usually crepuscular and/or nocturnal. Individuals in the Peruvian Pampas de Heath, an area uninhabited by man, are diurnal (Hoffmann *et al.*, 1975–1976, in Dietz, 1981). Maned wolves in Brazil are primarily nocturnal, hunting and traveling throughout the night, with occasional rest periods. During daylight hours they rest in thick cover and occasionally move short distances.

Captives generally conform to this activity pattern, showing nocturnal and crepuscular activity with crepuscular activity peaks (Dietz, 1984; Kleiman, 1972). They may also be active during late afternoon (Brady and Ditton, 1979), although this is perhaps a response to artificial feeding schedules. Kleiman (1972) observed that maned wolves in captivity were timid and fearful in the presence of humans, and they may have altered their activity patterns to avoid interacting with people. (The observations on which Kleiman's 1972 paper are based "were conducted at the Zoological Society of London from 1964 to 1966" and on observations made "at other zoos" (p. 792).)

REPRODUCTION

Females are annually monestrous. They bear litters of one to five pups after a gestation period of roughly 65 days (Brady and Ditton, 1979; Dietz, 1984; Kleiman, 1972; Faust and Scherpner, 1967; Langguth, 1975d). The largest litter recorded is one of seven pups, born in the São Paulo Zoo (Carvalho, 1976, cited in Dietz, 1984). Despite guidelines suggested by Brady and Ditton (1979), reproduction in captivity has a low success rate, and in many cases neonates are killed by their parents (Dietz, 1981; Faust and Scherpner, 1967).

Rasmussen and Tilson (1984) proposed that males may play a more important role in pup care than had previously been thought. Their observations on a captive family group suggest a higher level of parental care and tolerance than has been expected in a species usually considered solitary or asocial. The father of the captive litter they observed regurgitated food to his 5-month-old and older offspring. His offspring continued to successfully solicit regurgitations from him until the age of 10.5 months, when the pups were taken away by human keepers. Recent observations of another captive pair showed the same high level of male parental investment. The male participated in food provisioning, pup grooming, and family defense. The female of this pair was highly tolerant of the male's presence before, during, and after parturition (Bartmann and Nordhoff, 1984).

Little information is available on reproduction and parental care in free-ranging maned wolves. Anecdotal reports suggest that only very rarely are two adult wolves (i.e., both parents) seen in the wild with the pups (Dietz, 1984). Adult pairs appear to be monogamous, with a long-term pair bond (Bartmann and Bartmann, 1986).

Longevity in captivity is 12–15 years (da Silveira, 1968; Pithart *et al.*, 1986). In free-ranging individuals, parasites (particularly nematodes, which may destroy the kidneys), cystinuria (a potentially fatal inherited metabolic disorder), and human-caused deaths seem to be the most important factors contributing to mortality (Meritt, 1973; Dietz, 1984).

SOCIAL ORGANIZATION AND BEHAVIOR

The majority of the information on sociality and behavior in free-ranging maned wolves comes from Dietz's (1981, 1984) and Bartmann and Bartmann's (1986) field studies of radio-tagged individuals in the Serra da Canastra National Park, Minas Gerais, Brazil. This is a small area (715 km²), predominantly grassland, which includes areas of human settlement. Studies of captives have contributed the bulk of information on maned wolf behavior and social organization (Brady, 1981, 1982; Biben, 1983; Kleiman, 1972; Rasmussen and Tilson, 1984; Bartmann and Nordhoff, 1984).

The basic social unit in the wild is the mated pair, which shares a common home range. The male and female actually spend very little time in close association (Dietz, 1984; Bartmann and Bartmann, 1986). Pair members only rarely hunt or travel together, and probably never rest together (Dietz, 1984). Anecdotal reports from Brazil indicate that family groups of maned wolves are seen very rarely (Dietz, 1984), and groups of more than two adults have never been seen. The few other reports of maned wolf sightings confirm that they are usually seen alone (Dennler de la Tour, 1968, cited in Dietz, 1984). Dietz (1984) observed three exclusive home ranges occupied by mated pairs. These pairs "maintained strict territoriality"—i.e., territorial boundaries were almost never crossed by wolves from adjacent ranges and remained constant over a 2-year period.

Nomadic individuals travel the peripheries of occupied ranges. These nomads move in to fill vacancies created by death or abandonment of a territory. Areas left vacant by the disappearance of a maned wolf adult are appropriated by males from adjacent ranges. Males may play a larger role in territorial defense than females (Dietz, 1984).

Site-specific defecation (latrines) may play some role in territorial demarcation. Scats are often deposited in favored defecation sites, and the odor of maned wolf urine is quite strong (to human noses) along well-used trails, so urine marking may play some role in territorial demarcation as well. Garcia (1983) reported that urine marking was frequent and that the wolves marked vertical surfaces preferentially. At the National Zoological Park Conservation and Research Center, Front Royal, Virginia each maned wolf has two or three preferred sites where it deposits virtually all of its feces (Dietz, 1984).

Behaviors exhibited by captive adults, directed toward one another, are predominantly agonistic or indifferent. Affiliative behaviors are relatively infrequent, taking place most often in association with reproductive activities. Observations of captives (Dietz, 1984) suggest that although the social organization of maned wolves is relatively inflexible, relations between individuals of the mated pair change as a function of reproductive phase. Proestrus is characterized by distance-decreasing behaviors and a decrease in mutual avoidance. Anestrus is characterized by the highest levels of distance-increasing behaviors. Dietz (1984) observed that the outcome of interactions between pair members was generally predictable for each pair. No consistent gender-based dominance patterns were evident.

Interactions between a captive pair at the São Paulo Zoo were consistently agonistic, and there was a low rate of body contact and a high rate of mutual avoidance (Garcia, 1983). Other typical behaviors of captive adults include maintenance of different feeding areas and mutual threat behaviors (Langguth, 1975d). Unfamiliar same-gender individuals generally fight. Kleiman (1972) placed maned wolves at the extreme of the family Canidae in terms of nonsocial behaviors and solitariness.

In contrast, Brady and Ditton (1979, p. 171) reported a higher degree of mutual tolerance and affiliative interactions, stating that captive maned wolves "coexisted peacefully by avoiding each other." Animals that were kept together generally established dominant/subordinate relationships. Acosta (1972), reporting on a litter of captive pups, stated that by about 1 month of age a dominance hierarchy was being established among the pups. A mother-and-daughter pair that were kept together until the daughter was over 1-year old frequently engaged in mutual grooming. A male and female introduced at the age of 8 months "coexisted peacefully, interacted often in a friendly fashion, and rested in contact with one another" (Brady and Ditton, 1979, p. 172). Captive maned wolf pups continued resting in contact with one another until separated by human intervention at the age of 7 months.

Individuals who had formed pair bonds (as opposed to simply being placed in the same enclosure) also exhibited similar rates of affiliative behaviors. Thus the degree of familiarity of the animals has a bearing on the nature of their interactions. Enclosure size often plays a critical role in determining the levels of mutual toleration in captive canids. If it is not possible to maintain a minimum distance, particularly during periods when affiliative behaviors may not be naturally present, such as during nonreproductive periods, then agonistic behaviors will be overrepresented. There are also individual differences in rates of expression of affiliative behaviors. Perhaps these factors account for the differences observed in maned wolves.

There are fairly protracted periods of parental care (Rasmussen and Tilson, 1984; Brady and Ditton, 1979). Adults feed pups by regurgitation for up to 10.5 months. Biben (1983) studied the ontogeny of social behavior in two litters of captive maned wolves and gave a fairly detailed report on their development in comparison with bush dogs (*Speothos venaticus*) and crab-eating foxes (*Cerdocyon thous*), two other South American canids. Brady and Ditton (1979) recorded that the intensity of interpup interactions increases in 6–10-week-old pups and decreases thereafter.

The vocal repertoire of captive maned wolves has been analyzed by Brady (1981, 1982), who provided by far the most complete account of acoustic communication in this species. Roar barks are the primary long-range vocalization. These are emitted in bouts of 3–30, and appear to function in long-distance mutual location (Kleiman, 1972; Brady and Ditton, 1979; Dietz, 1984; Brady, 1981).

Dhole *(Cuon alpinus)*
Credit: Tom McHugh/National Audubon Society
Collection/Photo Researchers, Inc.

CHAPTER 6
Genus *Cuon*

Cuon alpinus: Dhole

Dholes are the only members of the monospecific genus *Cuon* Hodgson, 1838. Alternate common names include wild dog, Asiatic wild dog, Indian wild dog, red dog, red wolf, and chennai. These are highly social, pack-hunting, carnivores. A number of field studies have been completed in India (Johnsingh, 1982; Fox, 1984; Barnett, 1978), but almost nothing is known about the habits of dholes outside of this region. In India, dholes have long been poisoned and killed for bounty. Agricultural expansion and overgrazing have destroyed huge amounts of habitat (Krishnan, 1972; Fox, 1984; Cohen, 1978).

The species is classified as vulnerable by the IUCN, is listed on Appendix II of CITES, was classified as rare by the U.S.S.R. Ministry of Agriculture in 1978, and was listed as an endangered species by the U.S. Department of the Interior in 1980 (Nowak and Paradiso, 1983; Ginsberg and Macdonald, 1990). No accurate population estimates exist. Dholes are elusive, wary of humans, and generally difficult to study: Johnsingh's (1982) 5,000 field hours resulted in roughly 100 hours of observation time.

DISTRIBUTION AND HABITAT

The distribution of dholes is extremely broad, extending throughout eastern and central Asia. Dholes occur as far north as the Altai Mountains of the U.S.S.R., perhaps as far north as southern Siberia. From there, their range extends radially southward, encompassing Mongolia, much of China, Thailand, the Malay Peninsula, Sumatra, and Java. To the east, dholes occur in Tibet, Nepal, India, and possibly in Pakistan. They are not present in Sri Lanka or Borneo (Roberts, 1977; Johnsingh, 1982, 1985; Cohen, 1978; Pocock, 1936; Fox, 1984; Prater, 1965; Stains, 1975). Fossil remains indicate a distribution that was once even broader (Johnsingh, 1985). See Johnsingh (1985) for detailed distribution information on southern Asia. There are no reports of sympatry with other pack-hunting carnivores (Johnsingh, 1982). Much of the dholes' range does overlap with that of wolves (*Canis lupus*), but whether these two species ever interact or share home range is unknown.

Dholes occupy an enormous variety of habitats. In the northern reaches of their range in the U.S.S.R. they inhabit dense forests, river gorges, and mountainous alpine regions. Individuals have been trapped at 3,000 m. In winter they may migrate to less snowy zones (Novikov, 1962; Sosnovskii, 1967; Fox, 1984). In Nepal, they occur in alpine regions above treeline from 90 to 5,000 m (Mitchell, 1977). In Ladakh and Tibet they inhabit open country (Mitchell, 1977; Prater, 1965; Fox, 1984). In central Asia they usually inhabit forests and are occasionally found on the open steppes (Prater, 1965). Farther south the preferred habitat is scrubland and forest. In India they exist almost exclusively in dense forests and thick scrub jungles and are not found in open country (Krishnan, 1972; Cohen, 1978), perhaps because they have been forced to retreat due to human pressures from hunting. They occur in the hills of India at over 2,000 m, and dense montane forests at elevations of up to 3,000 m are the preferred habitat in Thailand (Lekagul and McNeely, 1977, cited in Nowak and Paradiso, 1983). Dens are burrows or rocky caverns (Cohen, 1978). In southern India, dens are found in rocky outcrops and the banks of dry creeks, obscured by dense vegetation (Fox, 1984).

The present range of dholes has been much reduced due to human activities. Dholes have become rare in or have disappeared entirely from many regions in central Asia, large parts of India, and eastern China (Müller-Using, 1975e).

PHYSICAL CHARACTERISTICS

The size and coloration of dholes varies regionally. On the whole, the dimensions are those of a small, slender wolf or a pariah dog. Individuals from the northern part of the distribution are larger than their southern conspecifics, and males' average weight is greater than that of females (Johnsingh, 1982; Prater, 1965; Cohen, 1977; Fox, 1984). Average weight of males is 15–20 kg, that of females 10–13 kg (Sosnovskii, 1967; Cohen, 1978). Head-plus-body length is 80–113 cm, height at shoulder is 42–55 cm, and tail length ranges from 40 to 50 cm

(Johnsingh, 1982; Novikov, 1962; Cohen, 1978; Mitchell, 1977; Prater, 1965). Pelage occurs in a variety of colors including tones of red, brick, mahogany, and light tawny. The sides and legs may be lighter in color, and the neck, shoulders, and upper parts of the head may be darker. Hairs on the back are sometimes black tipped. The underparts are paler, tawny to white, and some individuals have white throat patches that extend to the chest and underparts. The underfur is thick and light in color. As a general trend, the pelage is paler or tawnier in northern latitudes, becoming redder in southern parts of the distribution (Johnsingh, 1982; Mitchell, 1977; Novikov, 1962; Cohen, 1977, 1978; Sosnovskii, 1967; Fox, 1984). In winter, particularly in the northern parts of the range, the coat is long and furry with a dense undercoat. In some regions, the winter coat may be yellow-gray. In the spring, between March and May, this winter coat is shed for a shorter, coarser, and sparser summer one (Sosnovskii, 1967; Stroganov, 1962; Novikov, 1962; Clutton-Brock *et al.*, 1976).

The legs are long and slender, and fur covers the paws (Sosnovskii, 1967; Novikov, 1962; Stroganov, 1962). The tail is proportionately quite short relative to other canids, and it is well furred. It may be considerably darker than the rest of the coat and is usually black tipped, though some dholes have tufts of white or gray at the end of the tail (Novikov, 1962; Stroganov, 1962; Sosnovskii, 1967; Cohen, 1977, 1978; Prater, 1965; Fox, 1984; Burton, 1941). A well-developed caudal gland is present (Fox, 1984). The muzzle is deep and powerful, the facial region quite short. The ears are erect and rounded, and may be tipped in black fur. Their insides are well furred (Krishnan, 1972; Pocock, 1936; Fox, 1984; Stains, 1975; Novikov, 1962). Females have 14 or 16 mammae instead of the usual 10 found in *Canis* (Krishnan, 1972; Pocock, 1936; Cohen, 1978). These animals have a strong and distinctive odor (Burton, 1941).

The dental formula of *Cuon* differs from the usual formula in the Canidae. Dholes have only two true molars on either side of the lower jaw instead of the usual three. Thus the dental formula for *Cuon* is incisors 3/3, canines 1/1, premolars 4/4, molars 2/2 = 40. The usual pattern in the family Canidae is incisors 3/3, canines 1/1, premolars 4/4, molars 2/3 = 42. Relative to most members of the Canidae, the skull of *Cuon* is broad with a short rostrum, a trait that enables dholes to exert an extremely powerful bite (Novikov, 1962; Stroganov, 1962; Krishnan, 1972; Cohen, 1978; Mitchell, 1977).

TAXONOMY

Dholes are the only species in the genus *Cuon* Hodgson, 1838. They have long been placed in the subfamily Simocyoninae, along with African hunting dogs (*Lycaon pictus*) and bush dogs (*Speothos venaticus*). Recently it has become evident that the subfamily Simocyoninae is not a valid entity, and the designation is falling into disuse. As long ago as 1945, Thenius (1945, cited in Cohen, 1977) pointed out that similarities in dentition within the three species placed in this subfamily could be due to convergent or parallel evolution, and noted that other structural features are thoroughly different. Furthermore, these three monotypic genera seem to be more closely related to other genera than to each other. The analysis of all physical characteristics by Clutton-Brock *et al.* (1976) supports this. Müller-Using (1975e) stated that there are 11 subspecies, Stains (1975) stated that there are 10, and Cohen (1978) reported that two are currently recognized.

DIET

Dholes are carnivorous pack hunters. Throughout their range they rely primarily on medium-to-large ungulates, while also hunting smaller prey as well. All recent dietary studies have been conducted in India (Fox, 1984; Fox and Johnsingh, 1975; Barnett *et al.*, 1980). In the regions of India where dietary preferences have been studied, chital (small deer) are the primary prey. Dholes also hunt sambar, wild pigs, muntjacs, and mouse deer (chevrotains). Large packs attack buffalo (Fox, 1984; Barnett *et al.*, 1980; Prater, 1965). In Tibet and Ladakh, dholes hunt wild sheep and antelope, and in Kashmir, markhor, musk deer, and ghoral. In the U.S.S.R., medium-sized ungulates are the usual prey. Dholes hunt reindeer, wild sheep, goats, roe deer, badgers, and musk ox (Sosnovskii, 1967; Novikov, 1962; Cohen, 1978; Müller-Using, 1975e). Although domestic stock is common in much of the dhole's home range in India, where it is often allowed to roam freely, it constitutes only a very small fraction of the diet (Barnett, 1978; Fox, 1984; Krishnan, 1972). However, herders in Nepal complain about losses of livestock to dholes (Mitchell, 1977). Dholes have been seen killing panthers, bears, and tigers (Prater, 1965; Ollenbach, 1930; Khajuria, 1963; Connell, 1944), sometimes in disputes over kills. Smaller prey, primarily lagomorphs and rodents, are important and may be hunted by individuals or pairs of dholes. Dholes eat hares, palm squirrels, and field rats, and these assume greater dietary importance in the dry season in India (Barnett *et al.*, 1980; Fox, 1984). Other dietary items include birds, lizards, insects, and vegetable material, including grasses, leaves, fruit (particularly of the *Zizyphus*), and rhubarb (Barnett *et al.*, 1980; Fox, 1984; Schnitnikov, n.d., cited in Müller-Using, 1975e). Dholes may consume carrion, but not with any great frequency (Morris, 1937, cited in Davidar, 1975).

Dholes have a reputation as ferocious animals, due to their killing method, which, in the case of large prey, consists of pursuit and evisceration. This killing method, like that of African wild dogs (*Lycaon pictus*), has contributed to human hostility toward them. Despite their reputation as ferocious killers, there is no record of dholes killing a human being, and the single reported incident of an attempted attack did not result in injury (Burton, 1941; Krishnan, 1972; Fox, 1984).

ACTIVITY

Generally, dholes have a bimodal, crepuscular activity pattern. During the heat of the day they are usually inactive and resting, though in India during the cooler rainy season they may be active and hunt at any time of day. Occasionally they hunt during moonlit nights (Johnsingh, 1982; Krishnan, 1972; Fox and Johnsingh, 1975; Sosnovskii, 1967; Cohen *et al.*, 1978; Fox, 1984; Prater, 1965).

REPRODUCTION

The reproductive behavior of dholes reflects their highly social nature. Pup rearing is a social activity. Unlike most other canid species, mating is not confined to a narrow season. In India it occurs any time from September to January (Cohen, 1977, 1978; Johnsingh, 1982). Prater (1965) stated that the main breeding season in peninsular India is November and December. Captives in the Moscow Zoo mated in February (Sosnovskii, 1967). Gestation is usually 60–63 days (Cohen, 1978; Sosnovskii, 1967). In India pups are born most often in January or February (Prater, 1965). The average litter is 4 or 5; the maximum is 9 or 10 (Sosnovskii, 1967; Burton, 1941; Prater, 1965; Cohen, 1978). Up to 12 pups may live in a single den, but these are probably the offspring of two or more females. Females may den and rear their litters together (Cohen, 1977; Prater, 1965). Prater (1965) stated that a number of females may select a den site and form a breeding colony. According to Fox (1984), a clan may have several different-aged litters. There is some confusion on this point, however, since Johnsingh (1982) stated that breeding is restricted to only a single female within a pack, as is also the case for the highly social, pack-living African wild dogs (Lycaon pictus) and for wolves (Canis lupus). The entire pack helps to provision the young. While the pups are confined to the den area, pack members feed them, along with the lactating females. "Guards" — dholes that remain behind at the den site with the pups while the rest of the pack is out hunting—are also fed (Johnsingh, 1982; Fox, 1984). The genetic relationship of these guards to the pups and their parents is unknown. This type of guarding by nonparent pack members occurs in several other canid species as well, and is indicative of a high degree of sociality. At 70–80 days of age, the pups leave the den area. By the age of 7 months they are participating in prey killing with the other members of the pack (Johnsingh, 1982). According to Burton's (1941) account of captives raised by humans, dhole pups engage in agonistic behavior toward one another until they are 7 or 8 months old, by which time a dominance hierarchy is established. See Fox (1984) for information on development in free-ranging pups. Nothing is known about the dynamics of pup dispersal.

Sosnovskii (1967) described the development of captives at the Moscow Zoo. Captive breeding has occurred repeatedly and successfully in both the Moscow and Peking zoos (Sosnovskii, 1967; Müller-Using, 1975e). Fox (1984, p. 93) stated that no breeding stock exists in captivity except for a few individuals in the Moscow Zoo and Duisberg Zoo (Germany) "whose viability is questionable." Life span in captivity is 15–16 years (Sosnovskii, 1967).

SOCIAL ORGANIZATION AND BEHAVIOR

Dholes are highly social animals, comparable in the degree of their sociality to African wild dogs (Lycaon pictus) and wolves (Canis lupus). Dholes are group-living animals with a bilevel social organization. Packs are hunting and feeding units, and pack members remain together more or less constantly. Clans, formed by the assembly of two or more packs, are the larger level of social organization (Fox, 1984).

Clan assembly revolves around affiliative social interactions during rest and play (Fox, 1984). Clans coalesce and disperse in a loosely organized way. Nothing is known about the degree of genetic relatedness of clan or pack members. It is likely that packs are family units (Fox, 1984). Loners appear to be rare in dhole populations (Johnsingh, 1982). Average pack size is quite variable, ranging from three to 28 adults. Ramanathan (1982) observed a pack of 28, including both pups and adults, in Tamil Nadu, India. Johnsingh (1979, in Fox, 1984) observed a pack of 18, which split up during denning season. From seven to 11 dholes remained in the pack's core area, while the remainder dispersed. The mean number of adults in another pack was 8.3; including pups, pack size increased to 16. This pack rarely split into smaller groups (Johnsingh, 1982). Average pack size was five in another study in India, although during the dry season this decreased to an average of three (Fox, 1984). Occasionally, as many as eight–12 dholes were seen together. In Thailand, groups of more than three are seen only rarely (McNeely, personal communication, cited in Cohen, 1978). Adult sex ratios consistently favor males by a margin of 2:1.

Several packs may assemble after hunts, but only very rarely do groups this large hunt together (Fox, 1984). These clans break up readily into smaller groups for hunting, then later reassemble. Clans of 40 have been seen (Davidar, 1975), and some reports give figures as high as 100 or more. Large assemblages are more frequent during the hot weather and the rainy season (i.e., not during the breeding season) (Prater, 1965). During the end of the monsoon and at the beginning of denning season, dholes are most frequently seen in small groups (Fox, 1984). In recent years, sightings of large groups have been less frequent (Davidar, 1975). This may be due to declines in dhole populations or to changes in social organization brought about by habitat destruction or decline of the prey base. Behavioral adaptations to long-term, intense human persecution may play some part.

There is no clear evidence of dominance hierarchies within packs, although in the free-ranging pack observed by Johnsingh (1982) one adult male was clearly dominant over the rest of the pack. Displays of active submission were directed toward him by other adults in his pack. Aggressive interactions among adults are relatively rare (Davidar, 1975). Pack members are highly tolerant of one another at kills, and there is a low level of competition for access to kills (Fox, 1984). In captives at the Moscow Zoo, disputes often led to fights among same-sex individuals (Sosnovskii, 1967). Perhaps this aggression was due, at least in some part, to the enforced proximity of captivity. Free-ranging dhole pups and subadults also fight among themselves (Johnsingh, 1982; Davidar, 1975). This pattern of early intraspecific aggression is characteristic of animals that establish a dominance hierarchy among themselves during subadulthood (a period often characterized by intense fighting). Later, once a dominance hierarchy has been established, they settle into more peaceful relations and remain generally nonaggressive for as long as the hierarchy is not disrupted or challenged. Until recently, it was thought that no dominance hierarchy existed among African wild dogs. Only after long-term, detailed research on a single group, the sort of work that has yet to be done for dholes, did it become clear that a hierarchical social organization does exist. This may well turn out to be the case for dholes. Adults play with one another as well as with pups. The play behavior repertoire includes soliciting play with lowered forequarters, wagging the tail, and engaging in chase-and-ambush games (Johnsingh, 1982; Fox, 1984). Pups also play frequently among themselves (Johnsingh, 1982). As is true for all of the highly social species of canids, there is a prolonged association between parents and offspring (Prater, 1965). Davidar (1975, p. 119) raised a dhole pup and found it to be "extremely sociable, outgoing, and aggressive rather than cringing, but . . . disturbed by novel stimuli such as unfamiliar noises."

The home range of an Indian pack was 40 km², with an average population density of 0.35–0.9 individuals per km² in a 20 km² core area (Johnsingh, 1982). During the denning period, the area this pack used intensively was reduced to 11 km². No other estimates of the size of pack's home ranges are available. Dholes from northern regions may move vast distances, up to 600 km from their original home range (Müller-Using, 1975e). Whether such moves are related to dispersal or are the movements of "lone rangers" is unknown. Nothing is known about dispersal patterns, or about territorial or interclan interactions. Territorial marking behavior does not seem to be particularly well developed. Dholes have not been observed scent-marking along trails (Davidar, 1975), and Fox (1984) saw no evidence of dholes urine-marking their territories or travel routes. Dholes in captivity urine-mark one another, and they may perform hand stands to do so, like bush dogs (*Speothos venaticus*) and African wild dogs (*Lycaon pictus*) (Fox, 1984). Dholes practice site-specific defecation, and groups doing so produce communal latrines (Johnsingh, 1982; Fox, 1984; Cohen, 1977). These latrines may be located at conspicuous sites, such as the intersection of two trails, and dholes have a predilection for sites they have used before (Fox, 1984). It is not known whether these communal defecation areas, which may contain 40 or more dhole scats, serve as territorial markings or have an advertisement function. Fox (1984) averred that they probably do not, but stated that further study is needed. He observed no spatially or temporally consistent pattern of scent marking.

Dholes and pi dogs (pariah dogs, *Canis familiaris*) sometimes hunt together, but at kill sites dholes have complete priority of access (Davidar, 1975). Dholes have a wide variety of vocalizations, including the long-range whistle, which appears to function as a contact-seeking or assembly call (Fox, 1984).

Falkland Island Wolf *(Dusicyon australis)*
Credit: C.H. Smith, "Dogs I" in Naturalist's Library (W. Jardine, ed.), vol. 18. Edinburgh: Lizars, 1839.

CHAPTER 7
Genus *Dusicyon*

Dusicyon australis: Falkland Island Wolf

The Falkland Island wolf is extinct. The species occurred only on the Falklands, about 500 km off the eastern coast of Argentina. One of the first published references to canids in South America was Kerr's 1792 description of this large wild dog he called *Canis australis* (Berta, 1987). Charles Darwin reported that this wolf was the only indigenous mammal in the Falklands. It preyed on birds, perhaps also on seals. Other than that, little is known of its natural history or behavior. The last individual was killed in 1876 (Nowak and Paradiso, 1983). The extraordinary tameness of the Falkland Island wolves rendered them easy prey for humans; they were killed by fur traders from the late 1830s on, and then later by Scottish settlers raising sheep on the islands (Nowak and Paradiso, 1983).

The taxonomic affinities of this species are unclear, and pathetically few preserved relics remain. Given these limitations, Berta's (1987) cladistic study of South American canids is the most clearly reasoned and detailed available. She stated that the *Dusicyon* lineage divided into two modern genera, *Dusicyon* and *Pseudalopex*. Falkland Island wolves represented doglike, derived characteristics, as opposed to the more generalized, vulpine *Pseudalopex* species. The *Dusicyon* line is represented by the now extinct *D. australis* and *D. avus*, the latter known only from the fossil record. The pelage of *D. australis* was distinct from any of the mainland *Pseudalopex*. It was more rufous, the tail tip was white, and there were white markings on the muzzle and inner part of the limbs (Clutton-Brock, 1977). The dentition had some unique features, but the teeth more closely resembled the other South American foxlike species than those of *Canis* (Clutton-Brock, 1977). Furthermore, the skull lacked an interparietal crest (a feature commonly found in *Canis*) (Clutton-Brock, 1977). The five extant species in *Pseudalopex* are very distinct from *Dusicyon*, retaining primitive characteristics, with few derived ones (Brum-Zorilla and Langguth,

1980, p. 1043). For these reasons, the five foxlike South American Canidae species in *Pseudalopex* should be kept in a separate genus, rather than lumped in with *Dusicyon* or *Canis*, at least until further evidence prompts taxonomic revision.

Clutton-Brock (1977) proposed that the Falkland Island wolf was either a relic of a domesticated form of mainland South American canid or a feral hybrid descended from a domesticated *Canis* x *Dusicyon* cross. Falkland Island wolves may have been tamed by humans and subsequently introduced to the Falklands in the early Holocene. This provocative suggestion is based on a number of well-reasoned observations, not the least of which is that *D. australis* was the only indigenous mammal on the islands when the first Europeans arrived in the 1700s and 1800s. The shortest distance from continental South America to the Falklands (500 km) makes autonomous recent dispersal highly unlikely. And it is unlikely that a large carnivore survived as the sole mammal on the islands throughout the Pleistocene, without any other representatives of any other mammalian species. Furthermore, *D. australis* is unknown as a fossil (Berta, 1987). Clutton-Brock (1977, p. 1341) drew an analogy between the situation of Falkland Island wolves and that of dingoes (*Canis familiaris dingo*). Both show characteristics associated with domestication, such as white pelage markings, "a wide muzzle with large somewhat compacted teeth in the premolar region, and expanded frontal sinuses." And of course both share the proposed situation of introduction to an island environment by human agency—except that dingo populations have withstood intense human eradication efforts beginning from roughly the same era. Falkland Island wolves were not so fortunate.

Fennec Fox *(Fennecus zerda)*
Credit: Anthony Mercieca/National Audubon Society
Collection/Photo Researchers, Inc.

CHAPTER 8
Genus *Fennecus*

Fennecus zerda: Fennec Fox

Fennecs are the smallest members of the family Canidae. They adapt readily to captivity, and there are numerous reports on the behavior of captives. Courting behaviors in captive pairs (Gauthier-Pilters, 1967) and physical and behavioral development (Koenig, 1970) have been well described. Little is known, however, about the habits of free-ranging foxes.

Fennecs are hunted intensively by humans in the Sahara region, and their population has declined in some parts of northwestern Africa (Müller-Using, 1975c). Pups are taken from their dens, fattened up, and marketed as food for humans (Rosevear, 1974).

DISTRIBUTION AND HABITAT

Fennecs exist in the northernmost tier of African countries, from Morocco through Algeria, Tunisia, Niger, Libya, Egypt, and the Sudan (Rosevear, 1974; Meester and Setzer, 1971; Harrison, 1968). Only two occurrences outside the African continent have been confirmed, one in the Sinai and one in Kuwait (Nowak and Paradiso, 1983). The species probably has reasonably wide distribution in the Arabian region, although conflicting reports are given on its distribution in this area (Harrison, 1968; Clutton-Brock *et al.*, 1976; Ewer, 1973; Meester and Setzer, 1971; Rosevear, 1974; Nowak and Paradiso, 1983).

Well adapted to arid environments, fennecs exist in desert and sub-desert habitats. They require a soft substrate for burrowing and are found principally in sandy areas. Stable sand dunes provide ideal habitat (Rosevear, 1974; Meester and Setzer, 1971; Dorst and Dandelot, 1969).

PHYSICAL CHARACTERISTICS

Densely furred foot soles, enormous ears, and pale, thick fur all function as adaptations to extreme desert environments. Fennecs weigh no more than 2 kg, with a head-plus-body length of 30–40 cm, height at shoulder of 15–17.5 cm, and tail length of 16–24 cm (Rosevear, 1974; Dorst and Dandelot, 1969). The basic pelage color is pale cream, often with a light fawn, reddish, or gray cast to it, and the underparts and limbs are a pale buff or cream. There may be a darker line along the spine. Individuals from North Africa have a reddish-sand color to the upper parts of their limbs, while individuals from farther south are nearly white on this part of the body. Their pelage is very dense, long, and soft, a thick coat being advantageous for the extreme cold of desert nights. The tail is very well furred and has a conspicuous caudal gland spot and a black-brown tip. The ears are both broad and long (up to 15 cm), darker on the back, and white inside (Rosevear, 1974; Dorst and Dandelot, 1969; Gauthier-Pilters, 1967; Harrison, 1968; Müller-Using, 1975c; Clutton-Brock et al., 1976). Rosevear (1974) included data on physical measurements of fennecs from various regions.

Overall, fennecs are small and delicate, and the skull follows this pattern. The only remarkable features are the greatly enlarged auditory bullae (Harrison, 1968; Rosevear, 1974; Clutton-Brock et al., 1976). In other respects, the skull is typically vulpine (Clutton-Brock et al., 1976). The dentition is feebly developed, with the same dental formula as V. vulpes. The canines are quite small and slender (Harrison, 1968; Clutton-Brock et al., 1976). Rosevear (1974) remarked on the sharply cuspidate nature of the teeth, which probably facilitates insectivory.

TAXONOMY

Fennecus zerda (Zimmermann, 1780) is the sole species in the genus Fennecus Desmarest, 1804 (Rosevear, 1974; Meester and Setzer, 1971; Nowak and Paradiso, 1983; Harrison, 1968). Proponents of taxonomic revision in Canidae have lobbied for inclusion of the fennecs in Vulpes (Clutton-Brock et al., 1976; Wozencraft, 1989; Macdonald, 1984) or the subgenus Vulpes as Canis (Vulpes) zerda (van Gelder, 1978), but these proposed reclassifications have not attained wide acceptance. Rosevear (1974) cited a 1954 chromosome study of Matthey's, which indicated that the genus Fennecus more closely resembles the wolves (genus *Canis*) than the foxes (genus *Vulpes*). Thus the taxonomic relationships of this genus remain unclear. The karyotype of fennecs is $2n = 64$: $NF = 70$ (Wurster and Benirschke, 1968). No subspecies have been described (Rosevear, 1974; Meester and Setzer, 1971).

DIET

Fennecs are omnivorous, with insects figuring prominently in their diet. Free-ranging individuals feed on such small rodents as gerbils and jerboas, small birds, eggs, lizards, vegetable material (particularly tuberous or bulbous roots, which may be important sources of moisture), and various fruits (Rosevear, 1974; Meester and Setzer, 1971; Gauthier-Pilters, 1967; Dorst and Dandelot, 1969; Müller-Using, 1975c). In captivity, fennecs' omnivory is even more thoroughgoing. Captives eat fish, carrots, dandelion leaves, commercial puppy chow, and honey (Weiher, 1976; Müller-Using, 1975c). Fennecs are capable of killing animals larger than themselves. Gauthier-Pilters (1962) recorded an instance of captive fennecs dispatching a full-grown rabbit (bear in mind that fennecs seldom weigh more than 1.5 kg). Fennecs seem able to survive indefinitely without access to free water (Banholzer, 1976, cited in Nowak and Paradiso, 1983; Schmidt-Nielsen, 1964, cited in Ewer, 1973).

ACTIVITY

This is a nocturnal species, but fennecs may engage in some crepuscular activity as well (Rosevear, 1974; Gauthier-Pilters, 1967; Meester and Setzer, 1971; Dorst and Dandelot, 1969; Müller-Using, 1975c).

REPRODUCTION

The reproductive patterns of fennecs conform to those of most members of the Canidae, with the exception of female fennecs' ability to produce more than one litter in a year. If the first litter is lost, the mother may bear a second one 2.5–3 months later. In one case, a third litter was produced within a year (Koenig, 1970). A gestation period of 50 days is common in captives, with a range of 49–63 days (Gangloff, 1972; Sowards, 1981; Gauthier-Pilters, 1967; Ewer, 1973; Rosevear,

1974; Petter, 1957; Volf, 1957). Litter size ranges from one to six, with two to five usual in captives (Koenig, 1970; Weiher, 1976; Gangloff, 1972; Meester and Setzer, 1971; Dorst and Dandelot, 1969; Rosevear, 1974; Volf, 1957; Dulaney, 1981; Saint-Girons, 1962). Offspring attain adult size by the age of 9–11 months (Müller-Using, 1975c).

In captivity, the male parent plays a significant role in pup rearing and also provisions the female at the den both before and after parturition (Sowards, 1981; Gauthier-Pilters, 1967). The male also actively defends the den and den area until the pups are about 4 weeks old (Koenig, 1970). Gangloff (1972) listed conditions necessary for successful breeding of captives: The breeding female must be allowed uninterrupted visual, auditory, and olfactory contact with conspecifics. This is perhaps indicative of the degree of sociality of fennecs.

No data on the longevity of free-ranging fennecs is available. One captive lived 14 years 3 months, and another lived close to 13 years (Saint-Girons, 1971; M. K. Jones, Jr., personal communication, cited in Nowak and Paradiso, 1983). Fennecs are preyed on by jackals, hyenas, and domestic dogs (Gauthier-Pilters, 1967).

SOCIAL ORGANIZATION AND BEHAVIOR

Fennecs are moderately social. The basic social unit is the mated pair and their offspring. The offspring of a previous year may remain with the family group (Gauthier-Pilters, 1967), although this observation needs more substantiation (Bekoff, 1975). Several dens may be interconnected or located close together (Bueler, 1973). Captives exhibit relatively high rates of affiliative behaviors, such as resting in contact (sometimes resting piled on top of each other), and play. Play behaviors are common among pups, juveniles, and adults, though adults are less playful around the breeding season when males exhibit heightened levels of aggression (Rosevear, 1974; Gauthier-Pilters, 1967). Gauthier-Pilters (1966) recorded detailed observations of play behavior in fennecs and stated that both object-oriented play and conspecific-oriented play are common. Other affiliative behaviors include a typical canid greeting display, incorporating tail wagging, squeaking, a greeting face, and ducked posture. Individuals may roll over as well. This display is directed both toward conspecifics and familiar human keepers (Gauthier-Pilters, 1962). Serious quarrels do occur between pair members (Gauthier-Pilters, 1967), and individuals may fight over food items, or play may escalate into agonistic behavior. Although otherwise quite social, fennecs hunt singly (Meester and Setzer, 1971). Fennecs are readily tamed by humans (Vogel, 1962; Gauthier-Pilters, 1967), a trait characteristic of the more social canids and, indeed, of social species in general.

No information on territoriality is available for free-ranging fennecs. Gauthier-Pilters (1962) recorded that only males urine-mark environmental objects, and they do so more frequently during the breeding season. Captives practice site-specific defecation (as opposed to defecating at sites located throughout their enclosure in a seemingly random pattern). They may deposit urine or feces in a shallow depression they have scraped with their feet, afterwards covering the spot by pushing sand with their nose or scraping with their hind feet.

Short-range sounds include whines and purrs. Purring occurs in behavioral contexts similar to those in which domestic cats (*Felis cattus*) purr (Rosevear, 1974; Gauthier-Pilters, 1967). Fennecs and kit foxes (*Vulpes macrotis*) are the only members of the Canidae in which purring has been observed. Growls and snarls are emitted during agonistic interactions with conspecifics. Barks, similar to those of small domestic dogs, occur in threat and warning contexts (Rosevear, 1974; Koenig, 1970; Gauthier-Pilters, 1967; Harrison, 1968). Squeaking occurs as part of a greeting ceremony directed toward conspecifics or familiar human keepers (Gauthier-Pilters, 1967). Koenig (1970) mentioned a howl-scream accompanying reproductive activities.

African Wild Dogs *(Lycaon pictus)*
Credit: M.. Reardon/National Audubon Society
Collection/Photo Researchers, Inc.

CHAPTER 9
Genus *Lycaon*

Lycaon pictus: African Wild Dog

African wild dogs are the most highly social of the canids. They are also known as Cape hunting dogs, but this is a misnomer since their distribution includes most of Africa. Although they occur across a vast area, there are probably fewer than 7,000 individuals left. The species is classified as vulnerable by the IUCN although the IUCN/SSC canid specialist group recommends changing the listing to endangered (Ginsberg and Macdonald, 1990). Wild dog populations have undergone a precipitous decline due to human activities. They have been adversely affected, along with most other African wildlife, by the encroachments of human habitation on wild lands. The decline of wild ungulate populations has affected them as well (Kingdon, 1977). Outright killing by humans is also a key factor. Wild dogs are not particularly wary of humans, and they are often shot by hunters, farmers, and stockgrowers.

DISTRIBUTION AND HABITAT

The original range of African wild dogs encompassed an area from the southern edge of the Sahara to South Africa. The Sudan is the present northern limit of their distribution, which once extended into Egypt (Grzimek, 1975). From the eastern coastal countries of Ethiopia, Somalia, and the Sudan, the range sweeps westward through Mali, Niger, and the Ivory Coast. From there it extends to the eastern border of Guinea and Burkina Faso. African wild dogs also exist in Kenya, Tanzania, Zaire, the Congo, Zambia, Angola, Malawi, Botswana, Namibia, and South Africa (Kingdon, 1977; Rosevear, 1974; Meester and Setzer, 1971). It should be remembered that although the range is huge, the population is composed of fewer than 7,000 individuals. African wild dogs have vanished from many areas where they were once common and now exist only in remote or protected areas. See Childes (1988) for detailed data on distribution in Zimbabwe. There, African wild dogs occur only in protected areas. When they move out of these areas they are harassed or shot. In one area, pack sizes have declined by 99% in the period from 1980 to 1985. These reductions in pack size, an inevitable result of overall population decline, in turn affect population levels. Smaller packs are less efficient in defending their kills from hyenas (Childes, 1988), and fewer adult helpers at dens means lower rates of pup survival as well (Malcolm and Marten, 1982). In this downward spiral, decreasing population levels result in smaller pack sizes, which then result in decreased reproductive potential.

Wild dogs are found in a wide variety of habitats, including grasslands, savannas, and open woodlands (Kingdon, 1977; Rosevear, 1974; Meester and Setzer, 1971; Kruuk and Turner, 1967). They are seldom seen in dense forests (Dorst and Dandelot, 1969; Meester and Setzer, 1971). They are found on montane savannas, and a pack was sighted on the summit of Mount Kilimanjaro at 5,895 m (Wilson, 1975). Burrows, which are used only for three months each year during the breeding season, are the abandoned holes of ant bears, aardvarks, giant pangolins, or other diggers, which are appropriated and modified by the dogs (Reich, 1977; Rosevear, 1974).

PHYSICAL CHARACTERISTICS

Wild dogs are quite large. Shoulder height is 61–78 cm, head-plus-body length is 76–112 cm, and tail length is 30–41 cm (Kingdon, 1977; Rosevear, 1974; Grzimek, 1975). Weight varies a great deal, from 17 to 36 kg. East African individuals are smaller than their central and southern counterparts (Schaller, 1972; Estes and Goddard, 1967; Kingdon, 1977). Estes and Goddard (1967) stated that the weight of wild dogs from eastern Africa does not exceed 18 kg, while individuals from southern Africa are some 9 kg heavier and 7.6 cm taller. Pienaar (1973) stated that wild dogs from Kruger Park, South Africa, weigh from 22.5 to 27 kg. There is no sexual dimorphism (Frame et al., 1979).

The Latin word *picta*, from which the specific Linnaean name *pictus* is derived, means "painted" or "ornate" (Rosevear, 1974). This term aptly describes the patterned coat of these dogs. Sharply delineated blotches of black, tan, yellow, and white cover the body, chest, belly, and legs in a thin-haired coat. The coats of wild dogs from the eastern African steppe-savannas are generally darker than those of central and southern African individuals (Estes and Goddard, 1967). The patterning is remarkably variable, with no two coloration patterns the same (Rosevear, 1974; Kingdon, 1977; Dorst and Dandelot, 1969). Kingdon (1977) suggested that this distinctive coat pattern enables wild dogs to differentiate rapidly between conspecifics and prey during the often chaotic hunting and consumption of prey. It may also act as a unifying camouflage for the pack, allowing all pack members to blend together visually while interacting; this might serve an important social function in this extremely pack-oriented species. This group-affiliative function has been suggested for other conspicuously colored, highly social animals, such as zebras.

As is appropriate for animals that rely on high speed and long chases to catch their prey, wild dogs are slender and often look downright scrawny. Their muzzles are short, broad, and powerful, and are always black (Kingdon, 1977; Rosevear, 1974; Dorst and Dandelot, 1969). The ears are huge (11.5–12.5 cm long), rounded, and black on the back (Rosevear, 1974). Tails are short haired on the proximal end, tipped with conspicuous white tufts (Kingdon, 1977; Rosevear, 1974; Grzimek, 1975). These white markings, like those on dholes (*Cuon alpinus*), may help wild dogs keep visual contact with one another, particularly during a chase. Their legs are long, slender, and patterned with the same variegated coloration as the rest of the body. There are only four toes on each foot, as opposed to the usual five on the forefeet found on all other canids (Clutton-Brock *et al.*, 1976; Dorst and Dandelot, 1969; Rosevear, 1974). Females have 12–16 mammae (van Heerden and Kuhn, 1985); the usual number in *Canis* is 8–10. Wild dogs have a very strong musky odor, often described by humans as offensive, which seems to emanate from their entire body (Kingdon, 1977; Estes and Goddard, 1967; Shortridge, 1934). The skull is strongly built, and the dentition is massive. All the teeth are "relatively broad and strongly cusped" (Rosevear, 1974, p. 79; Clutton-Brock *et al.*, 1976). The auditory bullae are prominent but not large (Rosevear, 1974).

TAXONOMY

Lycaon was once placed in the subfamily Simocyoninae with *Speothos* (bush dogs) and *Cuon* (dholes) (Simpson, 1945; Stains, 1975). This classification, based on the common presence of a trenchant heel on the carnassials, has been discarded. *Lycaon* is now recognized as an anomalous species not closely allied with either *Speothos* or *Cuon*

(Clutton-Brock *et al.*, 1976). See Wayne and O'Brien (1987) for an interesting discussion of *Lycaon's* potential taxonomic affinities. The genus *Lycaon* Brookes, 1827 is monospecific (Clutton-Brock *et al.*, 1976). Four or five subspecies are recognized (Stains, 1975; Meester and Setzer, 1971), but see Rosevear (1974, p. 90) for comments on the validity of these subspecific divisions.

DIET

Wild dogs are carnivorous. They are cursorial hunters that catch their prey by outrunning it after a silent approach. The whole pack, except for young pups and the adults involved in pup care, participate in the hunt. The pack is essential to the wild dog's hunting success, since a single dog cannot bring down large prey. Wild dogs hunt primarily by sight (Estes and Goddard, 1967).

Prey varies from region to region, but wild dogs everywhere depend on medium-sized ungulates. Estes and Goddard (1967), who studied prey selection and hunting behavior on the Serengeti, found the prey breakdown to be as follows: Thomson's gazelles, 54%; juvenile and newborn wildebeest, 36%; Grant's gazelles, 8%; hartebeest, 2%. Kruuk and Turner (1967), also observing wild dogs on the Serengeti, found a similar diet. In western Africa, wild dogs subsist primarily on gazelle, reedbuck, and kob. They also prey on bushbuck, oribi, and duiker (Rosevear, 1974). Other prey include gemsbok, eland, warthog, zebra, ostrich, and young giraffe (Kingdon, 1977; Rosevear, 1974; Shortridge, 1934). Some packs on the Serengeti specialize in hunting zebra (Malcolm and van Lawick, 1975). Wild dogs also consume domestic stock, such as goats, sheep, and cattle (Schaller, 1972; Rosevear, 1974; Dorst and Dandelot, 1969; Roberts, 1951). This predictably brings human retribution on all wild dogs. In addition to large prey, they also hunt and eat smaller animals, but these they hunt individually and do not share with other pack members (Estes and Goddard, 1967). These include small mammals, such as ground squirrels, hares, springhares, cane rats, and other rodents, as well as birds (Rosevear, 1974; Kingdon, 1977; Dorst and Dandelot, 1969; Roberts, 1951). Carrion is consumed occasionally.

African wild dogs have been clocked during bursts of speed at 60–65 kph, and they are able to sustain speeds of 50–60 kph for considerable distances (Kruuk and Turner, 1967). Estes and Goddard (1967) reported that the average duration of the chases they witnessed was 3–5 minutes, covering 1.6–3.2 km. The success rate of wild dog hunts seems to be relatively high. Estes and Goddard (1967) observed an 85% success rate (50 kills), and Goodall (1970, cited in Rosevear, 1974) gave a figure of 43%. Kühme (1965a) reported that all the hunts he saw were successful, and Schaller (1972) gave a 70% success rate. In Kruger National Park, South Africa, hunting success rates are 70–90% (Mills, 1988). Prey is often disemboweled and partially eaten before it dies, a killing technique that has given the species a reputation as vicious killers. In fact, wild dogs consume their prey as rapidly as possible to protect it from other carnivores (hyenas, lions, jackals, and vultures) that are waiting to get some. The reputation of wild dogs as vicious, maniacal killers is undeserved, but surplus killing does occur (Kruuk, 1972b). The entire pack feeds simultaneously, without aggression (Reich, 1977). When pups are present, they are allowed to eat their fill before the adults feed (Reich, 1977). During the breeding season, wild dogs at a kill site will bolt their food, then return to the den to regurgitate food for the pups and for any adults who have remained behind with the pups (Kühme, 1965a,b). Free-ranging dogs eat about 2.7–5.0 kg per dog per day (Estes and Goddard, 1967; Rosevear, 1974). In captivity 1.5 kg per dog per day is adequate for maintenance (Dekker, 1968). They drink water when they can get it, although they can also go for long periods without it (Rosevear, 1974; Estes and Goddard, 1967).

ACTIVITY

African wild dogs usually hunt twice a day, and their hunting is crepuscular (Kingdon, 1977; Kruuk and Turner, 1967; Shortridge, 1934; Estes and Goddard, 1967; Kühme, 1965a,b; Pienaar, 1973; van Lawick-Goodall and van Lawick-Goodall, 1971). Periods of social activity and play bracket their twice-daily hunts (Estes and Goddard, 1967). Wild dogs may also hunt on bright, moonlit nights (Schaller, 1972; Kingdon, 1977; Estes and Goddard, 1967; van Lawick-Goodall and van Lawick-Goodall, 1971; Pienaar, 1973; Shortridge, 1934). Occasional hunting activity outside these times also occurs. If hunting is unsuccessful the pack may persist after darkness falls, even on nights that are not fully moonlit (Estes and Goddard, 1967). This does not happen often, especially since wild dogs seem to rely on vision for hunting. The usual crepuscular activity pattern coincides with peaks in the activity of many preferred prey species. All in all, wild dogs are active for only 4–5 hours a day (Kingdon, 1977; Schaller, 1972). The hottest part of the day is passed resting in the shade, when shade can be found (Kingdon, 1977).

REPRODUCTION

Usually only one female in a pack will breed and have pups in any given year (Malcolm and Marten, 1982; Frame *et al.*, 1979; Reich, 1977, 1981; Kingdon, 1977). The dominant female is usually the one to breed, and she mates with the dominant male, although subdominant males may copulate with her as well (Frame *et al.*, 1979). Changes in hormone levels resulting in suppression or delay of ovulation may be at least partially responsible for reproductive suppression in subdominant females (van Heerden and Kuhn, 1985). In Kenya, breeding may occur at any time of the year, and pups are born every month of the year except September (Frame *et al.*, 1979; Kingdon, 1977). In some areas in Kenya, there may be a breeding peak (Kingdon, 1977). In southern Kruger National Park, South Africa, breeding season is restricted to the dry season, when prey is concentrated at permanent water sources (Reich, 1981; Mills, 1988). Gestation ranges from 60 to 80 days (Kingdon, 1977; van Heerden and Kuhn, 1985). [Rosevear (1974), gave a range of 69–73 days.] From two to fifteen pups are produced, and average litter size is seven to ten (Frame *et al.*, 1979; Kingdon, 1977; Rosevear, 1974; van Heerden and Kuhn, 1985; Brand and Cullen, 1967). A female may have two litters in the same year if the first is lost soon after birth (Brand and Cullen, 1967). Kühme (1965b) observed a pack on the Serengeti in which two females had pups. Both females nursed all the pups and tried to steal them from one another. A similar situation has been described by van Lawick-Goodall and van Lawick-Goodall (1971). But the usual reproductive pattern is one litter per pack per year.

The remarkably large average litter size, almost twice that of closely related species, is linked to the social structure of this species: It is possible only because the reproducing pair monopolizes the energies of an entire pack (Frame *et al.*, 1979). Pup rearing is a social activity, and all the members of a pack contribute to the effort, all displaying nurturant behavior toward the pups, both feeding and guarding them

(Malcolm and Marten, 1982; Kingdon, 1977). Males are essential to successful pup rearing. When the mother of a 5-week-old litter died, four of the nine pups were successfully reared by the remaining pack members, who were all male (Estes and Goddard, 1967). There is a positive, though nonsignificant, correlation between the presence of helpers at the den and pup survival. These helpers are usually male (Malcolm and Marten, 1982). Past failures in captive breeding programs may be attributable in part to failure in recognizing the crucial role of the father and other pack members in pup rearing, as well as failure to provide adequate den sites (Cade, 1967; Brand and Cullen, 1967; Dekker, 1968). The pack and pups leave the den around the third month, but the pups are not proficient hunters until they are 12–14 months old (Frame *et al.*, 1979; Schaller, 1972). In a reverse of the usual mammalian pattern, it is the females who disperse and the males who remain with their natal pack (Wilson, 1975).

Wild dogs have few predators except man. Occasionally lions kill them (Kingdon, 1977). A number of canine diseases are responsible for some mortality. Maximum lifespan recorded for free-ranging individuals is 11 years (Frame *et al.*, 1979). Individuals in captivity live up to nine or ten years (Rosevear, 1974; Grzimek, 1975; Shortridge, 1934). See van Heerden (1986) for extensive observations on pathology among captives. Frame *et al.* (1979) provide detailed long-term information on reproduction and demographics for a free-ranging population on the Serengeti.

SOCIAL ORGANIZATION AND BEHAVIOR

This is the most highly social species in the Canidae. The pack is so fundamental to wild dogs' existence that it is always the basic social unit, and wild dogs are only rarely seen alone or in pairs (Dorst and Dandelot, 1969). A century or more ago, packs sometimes had 100 or more members (Rosevear, 1974; Shortridge, 1934). The writer Karen Blixen (Isak Dinesen) saw a single group of about 500 in Masailand (Kingdon, 1977). Maximum pack size has since decreased drastically, most probably due to human influences, and groups of 30 or more are now extremely rare (Rosevear, 1974). Whether this decrease in maximum pack size has brought with it changes in social organization and behavior will probably remain an unanswerable question. The majority of research on social organization has been conducted on the

Serengeti. Kruuk (1972b) gave an average pack size of 11.6, range 2–40 (n = 46), with only three observations of lone individuals. Frame *et al.* (1979) collected 10.5 years of continuous data there, and found that mean pack size is 9.8 (range 1–26) with a mean of 4.1 (range 0–10) adult males per pack and a mean of 2.1 (range 0–7) adult females per pack. (These figures for pack sizes exclude pups.) In most packs there is a preponderance of males (Childes, 1988; Rosevear, 1974; Malcolm and Marten, 1982; Frame *et al.*, 1979; Reich, 1981; Estes and Goddard, 1967). These skewed sex ratios are found in newly born litters (Pienaar, 1973; Malcolm and Marten, 1982; Estes and Goddard, 1967), and represent an interesting and as yet unexplained phenomenon.

Packs are composed of stable groups of genetically related males. Females disperse from their natal packs and breed elsewhere (Frame *et al.*, 1979). Upon reaching adulthood, most males remain behind with their male relatives, while no females stay with their natal pack after their third year (Frame *et al.*, 1979). Several sibling females may leave their natal pack and emigrate together to join another pack that lacks adult females (Fanshawe, 1989). Closely related dogs do not seem to mate with one another (Frame *et al.*, 1979), although Reich (1981) observed one instance of a young female replacing her mother as dominant and in subsequent years mating with her father, then with her older brother. Relations between pack members are consistently amicable. There is a high degree of mutual interdependence and extremely strong group cohesion (Schaller, 1972; Kingdon, 1977; Rosevear, 1974). No individual distance is observed, and pack members usually rest in close contact with one another (van Lawick-Goodall and van Lawick-Goodall, 1971; Wilson, 1975), a trait also shared among the canids by raccoon dogs (*Nyctereutes procyonoides*), bush dogs (*Speothos venaticus*), and bat-eared foxes (*Otocyon megalotis*). Packs cooperate completely in hunting and mutual defense (Estes and Goddard, 1967). In situations where competition would seem to be inevitable, participants assume appeasement postures, and there is no overt conflict (Schaller, 1972). Pack members rarely quarrel with one another, and serious fights are rare. Aggressive behavior does accompany female-female competition for care of the pups (Buitron 1977; Kingdon, 1977; Schaller, 1972; Estes and Goddard, 1967).

The extreme interdependence and cohesiveness in wild dog packs may be tied closely to the fact that every adult pack member constantly switches roles from provider to recipient and back again

(Kingdon, 1977). All members of a pack will regurgitate food for both pups and other pack members. Sick or disabled dogs who are unable to participate in hunting are provided for in this way (Wilson, 1975). The compulsion to share food is deeply entrenched; even 6-week-old pups regurgitate food for one another (Kühme, 1965a; Kingdon, 1977). Pack members share food even when there is not enough to feed anyone to repletion. Food sharing has assumed an important social role. Solicitations for food sharing (begging) have come to be used in appeasement contexts, and begging/appeasement gestures occur frequently in intraspecific interactions (Estes and Goddard, 1967; Kingdon, 1977).

Ritualized food-begging behaviors are a key component of the greeting ceremony, which is the most conspicuous social behavior of wild dogs. This ceremony involves the entire pack and occurs whenever the pack becomes active after a rest period and following the reunion of separated pack members (Kingdon, 1977; Estes and Goddard, 1967). It is important as a bonding activity and in motivating the pack for the upcoming hunt (Kingdon, 1977). In this ceremony, each pack member runs around frantically greeting the others. Face licking and poking the nose into the corner of another dog's mouth, gestures derived from food-begging behaviors, are prominent elements (van Lawick-Goodall and van Lawick-Goodall, 1971; Schaller, 1972; Estes and Goddard, 1967; Kingdon, 1977). Much vocalizing (twittering and whining) accompanies the ceremony. There is little discernible difference between the behaviors of high and low ranking individuals (van Lawick-Goodall and van Lawick-Goodall, 1971). Greeting ceremony behaviors grade insensibly into submissive behaviors. Thus, the importance of all-around mutual submission to the maintenance of the pack's social structure is evident. Early researchers suggested that no dominance hierarchy existed within packs (Kühme, 1965a,b; Estes and Goddard, 1967), although Estes and Goddard noted that in one pack a particular male was clearly the leader, and in another an adult female had this role. Subsequent studies suggest that clear dominance hierarchies, a separate one for each sex, are present (Reich, 1977, 1981; van Lawick-Goodall and van Lawick-Goodall, 1971).

African wild dogs are seasonally nomadic. Packs live and hunt in a small area of 2–5 km² during the 3 months that the pups are too young to travel. The rest of the year they range over an enormous area of up to 4,000 km² (Reich, 1977; Kingdon, 1977). Home ranges sizes are immensely variable, influenced at least in part by availability of prey.

They average 1,500–2,000 km², with as much as 50% overlap between the ranges of different packs (Frame *et al.*, 1979; Reich, 1977). In South Africa home ranges are smaller, roughly 450 km², with interpack range overlaps on the order of 10–20% (Reich, 1977). As would be inevitable for such indefensibly large ranges, territoriality is not well developed. Scent marking, so prominent in other canids, is not a conspicuous behavior in *Lycaon* (Wilson, 1975). However, the dominant or breeding female does urine-mark a great deal around the den while the pups are young (van Lawick-Goodall and van Lawick-Goodall, 1971). Population densities are related to prey base, water availability, and densities of lions and hyenas (Reich, 1981, cited in Childes, 1988).

Little is known about the relationships between packs, which may be friendly or antipathetic. Packs may simply ignore or avoid each other. On occasion, they may chase each other (van Lawick-Goodall and van Lawick-Goodall, 1971). Little is known about the processes involved in the formation of new packs (Kingdon, 1977).

Raccoon Dogs *(Nyctereutes procyonoides)*
Credit: Kenneth W. Fink/National Audubon Society
Collection/Photo Researchers, Inc.

CHAPTER 10
Genus *Nyctereutes*

Nyctereutes procyonoides: Raccoon Dog

Raccoon dogs are unlikely looking canids, with their stubby legs, raccoonlike facial masks, and pelage so thick it gives them a semispherical look. They are not closely allied taxonomically with any other members of the family Canidae. Although they are commercially important as furbearers, there have been few studies of their social organization or behavior.

DISTRIBUTION AND HABITAT

Raccoon dogs are indigenous to Japan, Manchuria, southeastern Siberia, China, and northern Indochina. From the late 1920s through the 1950s, 4,000–9,000 individuals were introduced into the western U.S.S.R. and the species has now spread through most of western Russia into Finland, Sweden, eastern Europe, and Germany, and in 1979 into France (Ikeda, 1986; Mikkola, 1974; Clutton-Brock *et al.*, 1976; Novikov, 1962; Stains, 1975; Ognev, 1962). Between 1935 and 1984 approximately 1.4 million km^2 were colonized in this natural expansion beyond the original introduction zones (Nowak, 1984). Individuals have recently been caught in England and France (Artois and Duchêne, 1982; Nowak and Paradiso, 1983). Their range is still expanding rapidly in places (Ward and Wurster-Hill, 1989). Raccoon dogs' relatively high reproductive rates, omnivory, overall adaptability, and tolerance of human presence have enabled this population proliferation. See Nowak (1984) for a bibliography on the distribution of raccoon dogs in Europe. However, raccoon dogs are now rare in some areas of Japan due to extreme human pressures, and populations in southeastern Siberia are also declining due to hunting and habitat destruction (Nowak and Paradiso, 1983), so there are limits to their adaptability.

Raccoon dogs exist in subarctic to subtropical climates. Preferred habitats are forest, forest borders, or thickly vegetated areas, often in river valleys. Areas bordering lakes and watercourses are favored. Thick cover provides small animals and vegetable material, mainstays of raccoon dogs' diet (Clutton-Brock *et al.*, 1976; Ikeda *et al.*, 1979; Nowak and Paradiso, 1983). Raccoon dogs rest and bear their pups in shallow burrows that have been abandoned by foxes or badgers; in hollow treetrunks; rock crevices; or other sheltered areas, such as dense vegetation (Ward and Wurster-Hill, 1989; Stroganov, 1962; Novikov, 1962).

PHYSICAL CHARACTERISTICS

Bulky, low-slung, with dense fur and distinctive facial markings, *Nyctereutes procyonoides* superficially resembles a raccoon more than a canid. Head-plus-body length is 50–68 cm, with a tail length of 13–25 cm. Summer weight is 4–6 kg, increasing to 6–10 kg before winter hibernation (Novikov, 1962; Stroganov, 1962; Mivart, 1890). The legs are very short: This is definitely not a cursorial animal. There is great variation in pelage coloration, and seasonal changes occur as well (Mivart, 1890), but the basic body colors are dusky brown to yellow-brown. The pelage is thick and soft with long guard hairs. Dorsal and tail guard hairs are black tipped over tawny thick underfur (Clutton-Brock *et al.*, 1976; Mivart, 1890; Nowak and Paradiso, 1983). Raccoon dogs have long been hunted for their thick fur. The limbs are blackish-brown to fawn in color (Clutton-Brock *et al.*, 1976; Mivart, 1890), and the chest and underbelly are brown to yellowish-brown or beige (Novikov, 1962; Clutton-Brock *et al.*, 1976; Mivart, 1890). The tail, very furry, is usually less than one third the total body length, and is blackish dorsally and a lighter yellow ventrally (Stroganov, 1962; Novikov, 1962; Mivart, 1890). The muzzle is fairly short, and distinctive facial markings resembling those of a raccoon (*Procyon lotor*) form a mask. The cheeks and areas surrounding the eyes are black, the sides of the neck are yellowish, and the chin and neck are brown. Social grooming is important in this species, and the distinctive facial markings may direct a groomer to the facial area (Kleiman, 1967). The ears are short, wide, and rounded with a white interior and brown margins (Mivart, 1890).

The teeth are small with the carnassial shear much reduced in comparison with the usual canid pattern. The molars are large and an extra upper molar is sometimes present, giving a dental formula of incisors 3/3, canines 1/1, premolars 4/4, molars 2 or 3/3 = 42 or 44 [as opposed to the standard canid pattern of incisors 3/3, canines 1/1, premolars 4/4, molars 2/3 = 42 (Ewer, 1973; Clutton-Brock *et al.*, 1976)].

TAXONOMY

Nyctereutes has no close affinities with any of the other members of the family Canidae. A "peculiar steplike subangular process for the insertion of the digastric muscle" is present, "a situation similar to that found in *Otocyon* [bat-eared fox] and *Urocyon* [gray fox] but different from *Canis, Vulpes, Alopex,* and *Fennecus*" (Stains, 1975, pp. 13–14). Another trait found only in *Nyctereutes* and *Otocyon* is an inverted U-shaped tail posture accompanying the expression of dominance, during attack, and accompanying sexual arousal (Kleiman, 1967). Recent biochemical analyses support the finding that *Nyctereutes* (along with *Otocyon* and *Urocyon*) is not closely related to any other canid taxa (Wayne and O'Brien, 1987). Berta's (1987) cladistic analysis found *Nyctereutes* and *Cerdocyon* to be primitive sister groups within the *Cerdocyon* clade. Five or six subspecies are recognized (Ikeda, 1986; Stains, 1975; Stroganov, 1962).

DIET

Raccoon dogs are omnivorous. True opportunists, they eat what's available. Their diets vary depending on season and locale, but everywhere small animals and plant materials form the mainstay of their diets. Vegetable materials, particularly important during the fall season, include all sorts of fruit, wild berries, and the seeds of grain crops such as oats (Ward and Wurster-Hill, 1989; Viro and Mikkola, 1981; Novikov, 1962; Stroganov, 1962). Small prey includes rodents, frogs and other amphibians, various birds including game birds and domestic fowl, and eggs (Viro and Mikkola, 1981; Novikov, 1962; Clutton-Brock *et al.*, 1976; Barbu, 1972). They also eat insects, mollusks, snakes, and lizards (Barbu, 1972; Ewer, 1973; Novikov, 1962). Along the seashore raccoon dogs eat crabs, sea urchins, and the carcasses of fish, birds, and marine mammals (Novikov, 1962). Carrion and human refuse are important during the winter months when other foods are scarce (Viro and Mikkola, 1981). Garbage is scavenged heavily in some areas of Japan (Ward and Wurster-Hill, 1989). In Japan, insects and plants, supplemented with fish and crabs, are the principal foods throughout the year (Ikeda *et al.*, 1979). In Finland in summer the most important diet constituents are, in decreasing order of importance, small mammals, plants, and amphibians. In winter, carcasses, small mammals, and plants predominate. Detailed information on the stomach contents of Polish raccoon dogs is included in Wlodek and Krzywinski (1986). Some raccoon dogs gain up to 50% of their body weight in the fall and then sleep through the winter months. Others who have not eaten enough will not hibernate, or will wake up and wander in search of food.

ACTIVITY

Raccoon dogs are usually reported as primarily nocturnal (Ikeda *et al.*, 1979; Müller-Using, 1975f; Mivart, 1890; Novikov, 1962; Nowak and Paradiso, 1983). A recent study in two regions of Japan showed regular diurnal, crepuscular, and nocturnal activity (Ward and Wurster-Hill, 1989). These are the only canids who hibernate, though "the process is neither profound for individuals nor universal for the species" (Nowak and Paradiso, 1983, p. 958). In the Far East, the hibernation period begins in November and continues until February, March, or April. Entire families of raccoon dogs hibernate in the same burrow (Stroganov, 1962). Individuals that are inadequately nourished may wander about in search of food (Novikov, 1962). Warm weather may interrupt hibernation as well, and in southerly regions of their range they may not hibernate at all (Novikov, 1962; Stroganov, 1962).

REPRODUCTION

Raccoon dogs mate in the early spring from January to March, with the copulatory tie lasting an average of 6 minutes (Valtonen *et al.*, 1977; Stroganov, 1962; Novikov, 1962). After a gestation period of 59–64 days, anywhere from three to eight pups are born (Valtonen *et al.*, 1977; Okuzaki, 1979; Novikov, 1962; Stroganov, 1962). Litters of up to 19 pups have been reported (Novikov, 1962; Stroganov, 1962). Raccoon dogs seem to be monogamous, although polygamy occurs among captives (Novikov, 1962). Males take an active role in caring for the pups. They provision their mates during late gestation and after parturition, and thereafter feed the pups (Ikeda, 1983, 1986; Stroganov, 1962). Ikeda (1983) described the behavior of a pair of wild-caught raccoon dogs and their captive-born offspring, and reported that the male assisted in the delivery of pups, fed the female and pups, and guarded the pups during absences of the female. Okuzaki (1979) described reproduction in captives.

Raccoon dogs are preyed on by wolves and other large carnivores, such as lynx, wolverines, and domestic dogs (Novikov, 1962; Stroganov, 1962). They are also eaten by humans in Japan, where they are considered to have an agreeable flavor, and where their bones are used in medicinal preparations (Nowak and Paradiso, 1983). They are hunted for their thick fur, which is used in both Japan and the U.S.S.R.

SOCIAL ORGANIZATION AND BEHAVIOR

Relatively little is known about the behavior of raccoon dogs. The basic social unit is the mated pair and their offspring of the year. Established pair bonds probably persist from year to year (Ikeda, 1986). Raccoon dogs never form packs (Ikeda, 1983). Some reports stated that raccoon dogs hunt in pairs or family groups (Clutton-Brock *et al.*, 1976), but a recent Japanese radiotelemetry study indicated that this is not consistently true (Ward and Wurster-Hill, 1989). Adults regularly sleep and rest in contact with one another, a behavior peculiar to the most social of the Canidae (African hunting dogs, *Lycaon pictus*, and bat-eared foxes or *Otocyon megalotis*) (Kleiman, 1967). Kleiman also pointed out that tail wagging as an expression of submission occurs in all canid species except for *Nyctereutes*. Home ranges vary from roughly 200 ha in European populations to 2.8 ha in Japan (Ikeda *et al.*, 1979; Nowak and Paradiso, 1983). In Japan home ranges on Kyushu were 49–59 ha (Ward and Wurster-Hill, 1989). Overlap in home ranges at these study sites indicated that raccoon dogs are not territorial.

Captives deposit feces at specific latrine sites, one per enclosure, used by both parents and offspring. Latrines are also used in the wild (Ikeda, 1983, 1986; Yamamoto and Hidaka, 1984). These serve important social functions and may be quite conspicuous (Ikeda, 1986). In the wild, raccoon dogs are not particularly wary of humans (Stroganov, 1962; Mivart, 1890), and they accustom themselves readily to captivity (Mivart, 1890).

Vocalizations include growls, whines, whimpers, and mews (Müller-Using, 1975f; Mivart, 1890; Stroganov, 1962; Kleiman, 1968). Raccoon dogs appear to be the sole representatives of the Canidae that do not bark.

Bat-Eared Fox *(Otocyon megalotis)*
Credit: Stephen J. Krasemann/National Audubon Society
Collection/Photo Researchers, Inc.

CHAPTER 11
Genus *Otocyon*

Otocyon megalotis: Bat-Eared Fox

Bat-eared foxes are small canids found in two separate regions of Africa. They are moderately social, but unlike any of the other moderately social members of the family Canidae, they are primarily insectivorous.

DISTRIBUTION AND HABITAT

The southernmost subspecies, *O. m. megalotis*, occurs from South Africa northward into Botswana, southern Angola, and western Rhodesia. *O. m. virgatus*, the northern subspecies, is found in Somalia, Ethiopia, and southern Sudan, and southward to Tanzania (Kingdon, 1977; Meester and Setzer, 1971; Ewer, 1973). Bat-eared foxes are adapted to arid or semi-arid environments. They are found in grasslands and savannas, along woodland edges, and in open acacia woodlands (Kingdon, 1977; Lamprecht, 1979). Free-ranging foxes in the Upper Limpopo Valley prefer bare, open habitat and require a short, sparse herbaceous layer with bare patches, as well as areas suitable for digging dens (Berry, 1978). In the Orange Free State, South Africa, and on the Masai Mara Game Reserve, Kenya, bat-eared foxes have a strong preference for short grass habitat (Mackie and Nel, 1989; Malcolm, 1986). Foxes in this region always occur within 2–3 km of open water (Berry, 1978). Dens are dug by the foxes themselves, or else are the modified burrows of aardvarks or aardwolves (Mackie and Nel, 1989).

PHYSICAL CHARACTERISTICS

Bat-eared foxes are relatively small canids. They range in weight from 2 kg to a little over 5 kg. Head-plus-body length ranges from 47 to 67 cm, and height at shoulder is 30–40 cm. Ear length is an extraordinary 11–13 cm (Kingdon, 1977; Müller-Using, 1975b; Malcolm, 1986). The basic pelage color is grizzled gray to buff with long guard hairs. The underparts and throat are light buff. The limbs are dark, shading to dark brown or black at their extremities. The muzzle and ears are blackish, and the insides of the ears are white. The tail is long, profusely bushy, and tipped in black (Kingdon, 1977; Dorst and Dandelot, 1969; Clutton-Brock *et al.*, 1976). Individuals of the eastern African subspecies, *O. m. virgatus*, tend toward a buff pelage with dark brown markings, as opposed to the black of *O. m. megalotis* (Smithers, 1966). Bat-eared foxes' proportionately large ears, a characteristic shared by many other inhabitants of hot, arid climates, may serve to disperse heat. They also help in locating prey.

In a sample of 25 males and 29 females from Botswana, females tended to be larger and heavier than males (Smithers, 1971, cited in Berry, 1978). Specimens from the Limpopo Valley (southern Africa) showed no significant sexual dimorphism (Berry, 1978). If there is sexual dimorphism with larger females, this would be a highly unusual, if not unique, attribute among the Canidae.

The dentition of *Otocyon* is unique among the Canidae. At least one additional molar is present on both the upper and lower jaws, giving a dental formula of incisors 3/3, canines 1/1, premolars 4/4, molars 3 or 4/4 or 5 = 46 or 50 (Clutton-Brock *et al.*, 1976; Ewer, 1973). (The usual Canidae pattern is incisors 3/3, canines 1/1, premolars 4/4, molars 2/3 = 42.) Bat-eared foxes have more teeth than any other heterodont placental mammal. In addition, the carnassials are much reduced, while the canines are large and "foxlike" (Clutton-Brock *et al.*, 1976). This dental specialization seems to be an adaptation to an insectivorous/omnivorous diet, providing an increased crushing surface and having an interlocking form that facilitates rapid chopping movements (Berry, 1978; Ewer, 1973). Bat-eared foxes have very rapid jaw movements while eating, "which result from muscles attached to an unusually developed sub-angular process on the lower jaw which accelerates jaw opening" (Gaspard, 1964, cited in Malcolm, 1986). The skull resembles that of the gray fox (*Urocyon cinereoargenteus*), a genus to which *Otocyon* is not closely related (Clutton-Brock *et al.*, 1976; Berry, 1978).

TAXONOMY

In the past, *Otocyon* was recognized as the only species in the subfamily Otocyoninae, one of the three subfamilies of the family Canidae

(Simpson, 1945). Recently this classification has been called into question, and it now seems well on the way to being discarded entirely (Langguth, 1971; Clutton-Brock *et al.*, 1976; van Gelder, 1978; Stains, 1975). Recent biochemical analyses support *Otocyon*'s unique generic status (Wayne and O'Brien, 1987): *Otocyon* does form a distinct lineage within the Canidae. Lamprecht (1979) stated that the genus is a link between *Protocyon*, a primitive fossil genus, and the present representatives of the Caninae.

Coetzee (1967, cited in Berry, 1978) recognized three subspecies: (1) *O. m. virgatus*, which is found from Tanzania northward to southern Sudan; (2) *O. m. megalotis*, which inhabits semi-arid regions of southern Africa; and (3) *O. m. canescens*, which exists in Ethiopia and Somalia. Meester and Setzer (1971) stated that *O. m. canescens* is probably the same subspecies as *O. m. virgatus*.

DIET

Bat-eared foxes rely primarily on invertebrate prey, but they also include vegetable material, small vertebrates, and carrion in their diets. Insects of various kinds—termites, beetles, locusts, ants, crickets, grasshoppers, spiders, millipedes, and scorpions—constitute the bulk of material eaten (Malcolm, 1986; MacDonald and Nel, 1986; Lamprecht, 1979; Nel, 1978; Kingdon, 1977; Schaller, 1972; Shortridge, 1934; van der Merwe, 1953a). This reliance on insect prey is unique among the Canidae. Bat-eared foxes also eat vegetable material, including berries, roots, bulbs, grasses, and fruit, as well as small vertebrate prey, such as rodents, lizards, snakes, and ground-nesting birds' nestlings and eggs. Mollusks are also consumed (MacDonald and Nel, 1986). A captive reared by humans was completely omnivorous, consuming toast, snakes, Marmite, bacon, chocolate cake, pudding, biscuits, mice, and insects (Smithers, 1966; Nel, 1978; Kingdon, 1977; Dorst and Dandelot, 1969; van der Merwe, 1953a; Turner 1968; Meester and Setzer, 1971; Lamprecht, 1978; Schaller, 1972; Berry, 1978; Malcolm, 1986). Carrion constituted 40% of the stomach contents (by volume) in two South African bat-eared foxes (Bothma, 1971a). Shortridge mentioned that these animals are found at carrion. Nel (1978) discussed diet composition and foraging habits of bat-eared foxes in the Kalahari Gemsbok National Park. Berry (1978) analyzed the stomach contents of 21 road-killed individuals from South Africa and Botswana and another 18 from the Transvaal (Berry, 1981), and described behaviors associated with food finding and feeding.

ACTIVITY

Crepuscular, nocturnal, and diurnal activity patterns have all been observed. Free-ranging animals in Botswana showed a consistent pattern of diurnal activity (Koop and Velimirov, 1982). They are commonly seen by day in the Kalahari region (Meester and Setzer, 1971). Kingdon (1977) stated that they are crepuscular. A captive reared by humans was crepuscular, consistently sleeping through the night (Turner, 1968). Bat-eared foxes in the Limpopo Valley are almost entirely nocturnal (Berry, 1978). Lamprecht's (1979) observations of bat-eared foxes on the Serengeti showed that 85% of activity occurred after dark, although individuals foraged on cool afternoons as well. In the Kalahari Gemsbok National Park, activity patterns are seasonally variable, with foxes chiefly diurnal in winter and nocturnal in summer (Nel, 1978). On the Masai Mara Game Reserve, Kenya, bat-eared foxes foraged almost entirely at night (Malcolm, 1986).

REPRODUCTION

Females are annually monestrous, with a gestation period of 60–75 days. Breeding generally occurs from August to October, although under certain conditions it may be aseasonal (Malcolm, 1986). From one to five pups are born. Six out of 12 dens dug up in the Limpopo Valley contained pups, with litters ranging from one to four (Berry, 1978). Offspring attain adult size by about 6 months (Kingdon, 1977; Rosenberg, 1971; Masopust, 1986; Berry, 1978; van der Merwe, 1953a; Dorst and Dandelot, 1969; Müller-Using, 1975b). There is a long-lasting pair bond; adults may renew pair bonds from year to year, returning to the same den site together (Mackie and Nel, 1989). The male parent may play a large part in pup rearing (Nel, 1978). Males stay with the pups at the den while the female is away foraging (Malcolm, 1986). Some variability in social structure associated with pup rearing may be present. Van Lawick-Goodall and van Lawick-Goodall (1971) recorded an instance of a social unit consisting of one male, two females, and five pups. The pups were all suckling from both the females.

Maximum lifespan is about 10 years in captivity, and one captive lived almost 14 years (M. Jones, personal communication, cited in Nowak and Paradiso, 1983). In the wild, sources of mortality include predation by leopards, hyenas, black-backed jackals, and large raptors. Humans also kill bat-eared foxes for their fur and for food (Berry, 1978; Dorst and Dandelot, 1969; Kingdon, 1977; Lamprecht, 1979; Shortridge, 1934; Davis, 1980; Müller-Using, 1975b).

SOCIAL ORGANIZATION AND BEHAVIOR

Bat-eared foxes are moderately social canids. The basic social unit is the mated pair, or the mated pair plus offspring (Lamprecht, 1979; Nel, 1978; Kingdon, 1977; Meester and Setzer, 1971). Bat-eared fox groups often split up to forage, and may hunt in groups of two, with separated subgroups moving through the same general area (Nel, 1978). Small family groups (two to five foxes) may forage closer to each other, often feeding less than 1 m apart (Nel, 1978). In Botswana, bat-eared foxes always forage in groups of three or four (Koop and Velimirov, 1982), and Smithers (1966) mentioned groups of five to eight individuals feeding together. On the Masai Mara Game Reserve, Kenya, groups ranged in size from one to nine (Malcolm, 1986). In the Kalahari Gemsbok National Park, observations over a 7-year period gave a mean group size of 2.72 (range 1–10, n = 623) (Nel *et al.*, 1984a). There, rainfall seemed to influence prey availability, which, in turn, was correlated with increased group size of bat-eared foxes. In remote areas, bat-eared foxes may form loose packs (Mills, personal communication, cited in Berry, 1978); it is unclear whether these groups are merely aggregations or more tightly knit associations. Their value in antipredator group defense should not be overlooked: Foxes in these large groups are able to drive off predators as large as leopards (Mills, personal communication, cited in Berry, 1978). Groups of bat-eared foxes have been known to mob black-backed jackals (Malcolm, 1986; Davis, 1980).

Bat-eared foxes show many affiliative behaviors. Mutual grooming, sleeping and resting in contact, and social play among adults and juveniles are all common (Lamprecht, 1979; Berry, 1978; Kleiman, 1967). Allogrooming is a frequent social behavior within bat-eared fox groups, particularly during early evening and after feeding (Lamprecht, 1979; Berry, 1978). Social play among adults without pups seems to be more common than in other canid species (Lamprecht, 1979), and play behaviors among juveniles are also prominent (Berry, 1978). Lamprecht (1979) observed periods of social interaction preceding foraging in the evening and following the group's return to the den in the morning. There is a low degree of intraspecific agonistic interaction. Bateared foxes fight over food items very infrequently (Nel, 1978; Lamprecht, 1979), perhaps because it is not energetically efficient to defend typical small food items. Lamprecht (1979) suggested that foraging in groups may be an adaptation to facilitate exploitation of resources whose exhaustion is time-dependent (i.e., insect aggregations) as opposed to those whose exhaustion depends on consumption.

The degree of site affiliation or territoriality varies regionally. In the Kalahari Gemsbok National Park, the home ranges of bat-eared foxes overlap extensively, and the foxes show no territorial defense or marking behaviors. In this area Nel (1978) observed four different groups foraging within the same 0.5 km². Nel found that over time, foraging areas tended to shift as well. In Botswana, foraging areas of individuals and groups overlap, and there is no evidence of intolerance between different groups (Koop and Velimirov, 1982). On the Masai Mara Game Reserve, Kenya, there is little intergroup aggression, and the vast majority of interactions between foxes from different groups are neutral or amicable (Malcolm, 1986). Five or six foxes may forage through a single area of 0.25 km² without interacting (Malcolm, 1986). On a game farm in the Orange Free State, South Africa, home ranges of different groups overlapped by as much as 55%, but there was no territorial defense (Mackie and Nel, 1989). This low level of intergroup aggression is unique among the Canidae. Berry (1978) found that several dens may occur within 100 m of one another in the Limpopo Valley, and on the Masai Mara, dens are clumped (Malcolm, 1986). Population density on one study area was 0.8–0.9 per km², except during May–August, when foxes gathered on short grass areas, and densities as high as 6 per km² were recorded (Malcolm, 1986). In contrast, on the Serengeti bat-eared foxes maintain home ranges, and resident pairs or groups are intolerant of nongroup conspecifics (Lamprecht, 1979). Estimated home range size is 0.25–1.5 km². Outside the breeding season in this area, bat-eared foxes form transient groups and leave their home ranges (Lamprecht, 1979).

Marking behavior is not particularly well developed. Unlike many canid species, bat-eared foxes do not show intensified marking behavior after agonistic encounters with conspecifics. On the Serengeti at least, individuals do mark the boundaries of, and sites within, the home range (Lamprecht, 1979). Kingdon (1977) recorded that males may urine-mark around the den; females may do so as well, but only during estrus. Bat-eared foxes, both captive and free ranging, practice site-specific defecation (Berry, 1978; Kingdon, 1977).

Both Turner (1968) and Smithers (1966) reared semi-tame bat-eared foxes. The animals adapted fairly well to conditions of captivity and had well-developed affiliative social interactions with their human keepers.

Lamprecht (1979) described the vocalizations of free-ranging bat-eared foxes, and provided several sonagrams. Components of the vocal repertoire are, with very few exceptions, of relatively low amplitude and seem to function at short range.

Chilla *(Pseudalopex griseus)*
Credit: Jeff Foott

CHAPTER 12
Genus *Pseudalopex*

Pseudalopex culpaeus: Culpeo

Common names for this species include culpeo, colored fox, large fox, culpaeo fox, colpeo fox, and Andean wolf (Fuentes and Jaksić, 1979; Cabrera, 1931; Osgood, 1943; Crespo, 1975; Stains, 1975; Langguth, 1975b; Clutton-Brock *et al.*, 1976). The name culpeo appears to be derived from the Chilean word *culpem* meaning "madness" or "folly," which refers to culpeos' lack of wariness and habit of exposing themselves to be shot by hunters (Osgood, 1943). The species has long been hunted, and its original habitat has been severely disturbed by the introduction of livestock ranching. Culpeos are now scarce and rarely seen (Fuentes and Jaksić, 1979; Jaksić *et al.*, 1980). *P. culpaeus* is on Appendix II of CITES (Nowak and Paradiso, 1983). On a more hopeful note, where food is abundant, these foxes are able to reproduce rapidly (Langguth, 1975b).

DISTRIBUTION AND HABITAT

Culpeos occur all along the western coastal region of South America from southern Colombia and Ecuador through Peru, western Bolivia, Chile, and Argentina, down to Tierra del Fuego (Berta, 1987; Langguth, 1975b; Crespo, 1975; Hershkovitz, 1957; Nowak and Paradiso, 1983; Ewer, 1973; Stains, 1975). They live in a range of habitats from arid to semi-arid, including the steppes of Patagonia, the deserts of the Tierra del Fuego region, the savannas of Ecuador, and the mountainous regions of the Andes up to 4,500 m or more (Crespo, 1975; Langguth, 1975b; Nowak and Paradiso, 1983). They seem to prefer mountainous habitats, at least in the northern part of their range (Fuentes and Jaksić, 1979).

Some individuals exhibit patterns of seasonal altitudinal migration. They move up to higher elevations in summer and in autumn descend to lower wintering grounds, a migration of 15–20 km. This altitudinal movement is tied to the seasonal movements of sheep and hares. However, many culpeos remain in either high or low country year-round and do not migrate (Crespo, 1975).

PHYSICAL CHARACTERISTICS

This is the largest of the five species in the genus *Pseudalopex* and, among all the South American canids, is second in size only to the maned wolf (*Chrysocyon brachyurus*). Head-plus-body length ranges from 52 to 120 cm, the tail is 30–51 cm, and weight ranges from 4 to 13 kg with an average of 7.35 kg (Langguth, 1975b; Crespo, 1975). See Fuentes and Jaksić (1979) for a discussion of latitudinal size variation in this species. The shoulders and back are gray with agouti guard hairs and fawn underfur. The sides of the body are paler than the back, and their underparts are buffy, grayish, white, or tawny (Langguth, 1975b; Clutton-Brock *et al.*, 1976; Hershkovitz, 1957; Osgood, 1934). The flanks and legs may be tawny or rufous, and the upper side of the feet is lighter in color (Osgood, 1943; Clutton-Brock *et al.*, 1976; Langguth, 1975b). The head, neck, and ears are tawny, rufous, or ocher. The chin is light tawny to ocher, and the jowls and lips are whitish or buffy (Osgood, 1943; Hershkovitz, 1957; Clutton-Brock *et al.*, 1976; Bueler, 1973). The tail is bushy and long, and may be more than half the length of the head and body. On its upper side it is grayish and on the underside, dull tawny. It is black tipped (Langguth, 1975b; Osgood, 1943; Hershkovitz, 1957; Clutton-Brock *et al.*, 1976). Culpeos are hunted for their fur (Osgood, 1943).

The skull is very similar to that of the side-striped jackal (*Canis adustus*), differing only in the relative sizes of the first and second upper molars. The canines and premolars are simple and foxlike (Langguth, 1975b; Clutton-Brock *et al.*, 1976). Culpeos are quite similar to pampas foxes (*P. gymnocercus*) in pelage and skull characteristics, as well as in general appearance.

TAXONOMY

Often placed in the genus *Dusicyon* as *D. culpaeus*, the species has also been placed in the genus *Canis* (Clutton-Brock *et al.*, 1976; Hershkovitz, 1957; Osgood, 1943; Stains, 1975; Langguth, 1975b; van Gelder, 1978; Iriarte *et al.*, 1989; Simonetti, 1986). Until the taxonomy of this and the other South American canid species is clarified, it should be retained in the genus *Pseudalopex* Burmeister, 1856 (Berta, 1987). Somewhere between four and six subspecies are recognized (Cabrera, 1931; Osgood, 1943; Stains, 1975). The Santa Elena fox (*D. culpaeolus*) may actually be conspecific with the culpeo. See Wayne *et al.* (1989) for details on morphological differences between *P. culpaeus* and *P. griseus*, particularly in the limited region of their sympatry. They include an interesting discussion of sympatry and morphological divergence among canid species as well.

DIET

Culpeos are generalist predators. Rodents and European rabbits are their primary prey (Iriarte *et al.*, 1989; Simonetti, 1986, 1988; Jaksić *et al.*, 1980, 1981, 1983; Meserve *et al.*, 1987; Crespo, 1975). Other prey items are bird eggs, insects, and snakes. Plant material and berries are important in the diet, particularly in summer and fall (Iriarte *et al.*, 1989)—a finding that contradicts Crespo's (1975) statement that the species is strictly carnivorous. The dietary study by Jaksić *et al.* (1983) in continental and insular Chile, based on both scat and stomach content analyses, showed a diet composed primarily of mammals, both rodents and lagomorphs, as well as birds and carrion. The culpeos also ate invertebrates and plant material. In northern Chile, rodents are the primary prey, and reptiles and birds are of secondary importance (Medel and Jaksić, 1988). Crespo's study of stomach contents (n = 96) in the Neuquén province of Argentina (all seasons) showed strict carnivory. The diet composition there was: rodents and hares, 61.4%; domestic mammals (sheep, cattle, and horses), 27.4%; other small items, such as small birds, 6% (percentages are by frequency of occurrence). Altogether, Crespo identified 17 different items. In Coquimbo Province, central Chile, a scat study showed almost exclusive reliance on small mammals (Meserve *et al.*, 1987). Another scat study in Coquimbo Province, which did not distinguish between *P. culpaeus* and *P. griseus* scats, showed mammals (primarily rodents and lagomorphs), 85.7%; reptiles, 7.3%; and birds, 6.7% of vertebrate remains. Insects, primarily Lepidoptera and Coleoptera, composed

95.8% of invertebrates consumed (Duran *et al.*, 1987). Culpeos also eat lizards (Langguth, 1975b; Bueler, 1973). Hershkovitz (1957) observed culpeos hunting rabbits, and stated that they prey on sheep. The establishment of sheep ranching and the introduction of European rabbits both occurred around 1915. These introductions have had a significant and long-term impact on the diet of culpeos. These foxes play an important role as a predator, especially in regulating the population of introduced European rabbits (Crespo, 1975). Sheep ranchers kill culpeos on the grounds that they pose a menace to their sheep (Langguth, 1975b).

ACTIVITY

Diurnal, crepuscular, and nocturnal activity all occur (Jaksić *et al.*, 1980, 1981). Hunting activity in one Chilean population was 51.5% diurnal, 31.6% crespuscular, and 16.9% nocturnal (Jaksić *et al.*, 1981). Culpeos' primary prey in this region is diurnal or crepuscular (Jaksić *et al.*, 1981).

REPRODUCTION

As is the case for almost all females of the family Canidae, female culpeos are annually monestrous (Crespo, 1975). Proestrus occurs between October and July, and estrus occurs between August and October. Gestation is 55–60 days, and the pups are born between October and December. Females bear about five young (range three to eight) (Medel and Jaksić, 1988). Dens are in low shrubs or between rocks (Langguth, 1975b; Crespo, 1975).

SOCIAL ORGANIZATION AND BEHAVIOR

Almost nothing is known about the social organization of culpeos. In Neuquén Province, Argentina, a population was 59.2% male, 40.8% female, with a population density of 1 fox per 140 ha. Age classes were biased heavily toward young foxes, with only 5% of the population over 24 months of age (Crespo and De Carlo, 1963, cited in Medel and Jaksić, 1988). Crespo (1975) stated that home range size is 4 km in diameter, although surely this must vary a great deal with differences in habitat, resources, and population levels.

Pseudalopex griseus: Chilla

Common names for this species include Chico gray fox, Argentine gray fox, Argentine fox, little gray fox, pampa fox, and chilla (Osgood, 1943; Clutton-Brock *et al.*, 1976; Stains, 1975; Bueler, 1973).

DISTRIBUTION AND HABITAT

Chillas have been heavily hunted by man and are now scarce throughout much of their range. They exist in the southern reaches of Chile and Argentina, below 25 degrees south latitude (Ewer, 1973; Clutton-Brock *et al.*, 1976; Nowak and Paradiso, 1983). They are scarce in Chile, where their original habitat has been severely disturbed and where they have been subjected to long-term (now illegal) hunting pressure (Duran and Cattan, 1985; Fuentes and Jaksić, 1979; Jaksić *et al.*, 1980). They are rarely found in the mountains of the Andes and seem to prefer habitats at lower elevations (Simonetti *et al.*, 1984; Fuentes and Jaksić, 1979; Stains, 1975). They are found on the plains and in the low mountains of Chile, Argentina, and Patagonia (Cabrera, 1931, 1958, cited in Clutton-Brock *et al.*, 1976); in the foothills and lowlands of Chilean coastal regions (Simonetti *et al.*, 1984); in open grasslands; on ocean beaches (Osgood, 1943; Greer, 1965, cited in Fuentes and Jaksić, 1979); and at forest edges (Nowak and Paradiso, 1983). Chillas were successfully introduced to Tierra del Fuego in 1950 with the goal of controlling rabbit populations (Duran and Cattan, 1985).

PHYSICAL CHARACTERISTICS

This is the smallest member of the *Pseudalopex* genus. Total body length is 80–90 cm, with a tail length of 30–36 cm (Osgood, 1934). Only a little shorter in body length than the pampas fox (*P. gymnocercus*), chillas are quite slender. The basic pelage color is gray with agouti guard hairs and pale underfur. The pelage coloration is quite close to that of the Sechura fox (*P. sechurae*) and the hoary fox (*P. vetulus*), but with a slightly more reddish cast. The underparts are pale gray. The head is rust colored and may be flecked with white, and the chin and base of the muzzle are black. The ears are large. The legs are pale tawny to reddish-brown, and the feet are tawny as well. There may be a transverse patch of black on the thighs (Osgood, 1934). The tail, moderately long and bushy, is a mixed pale tawny and black color on its underside (Osgood, 1934; Clutton-Brock *et al.*, 1976). The skull is foxlike, as are the widely spaced teeth. Overall, the skull is very similar to the skull of the culpeo (*P. culpaeus*), except the skull of *P. griseus* is smaller (Clutton-Brock *et al.*, 1976). Fuentes and Jaksić (1979) have documented size differences, which covary with latitude, in this species and in the culpeo.

TAXONOMY

The chilla has been placed in the genus *Dusicyon* (Clutton-Brock *et al.*, 1976; Fuentes and Jaksić, 1979; Jaksić *et al.*, 1980, 1983; Stains, 1975) and in the genus *Canis* (Duran *et al.*, 1987; Duran and Cattan, 1985; van Gelder, 1978; Langguth, 1975b). Osgood (1943) stated that there are three subspecies; Stains (1975) stated that approximately seven are recognized. Stains (1975) included Darwin's fox [the Chiloé Island fox (*Dusicyon fulvipes*)] with *P. griseus*, the species to which it seems most closely allied. Nowak and Paradiso (1983) do not recognize *D. fulvipes* as a species at all.

Darwin's fox [*Pseudalopex fulvipes* (Martin, 1937) or *Dusicyon fulvipes*], was originally supposed to be limited to Chiloé Island, Chile. Until recently, it had generally been accepted as an insular subspecies of *P. griseus* (Medel *et al.*, 1990). However, in the 1970s and 1980s a population of foxes was discovered on mainland Chile, 600 km north of Chiloé Island in Nahuelbuta National Park. It appeared to be sympatric with, and distinct from, the larger, lighter colored *P. griseus* (Medel *et al.*, 1990). These foxes were not abundant, but since 1986 their population appears to be increasing, concurrent with a chilla population decline (Jaksić *et al.*, 1990). Some other differences between the putative species have been observed: *P. fulvipes* is crepuscular and prefers forested habitat, where *P. griseus* is nocturnal and prefers open habitat (Jaksić *et al.*, 1990; Medel *et al.*, 1990). Of course the possibility remains that *D. fulvipes* is a variant, melanistic form of *P. griseus*. As Medel *et al.* (1990, p. 76) note, "the disjunct geographical distribution apparently shown by [*P. fulvipes*] is very puzzling." See Medel *et al.* (1990) and Jaksić *et al.* (1990) for scat analyses.

DIET

Simonetti *et al.* (1984) studied the diet composition of chillas in a coastal desert zone in northern Chile. From 121 fecal samples collected in mid-winter, they determined that the diet is composed primarily of rodents, secondarily of lizards and birds. Tenebrionid beetles make up the bulk of the invertebrate prey eaten. Chillas consume other invertebrates as well, but not in great quantities. The proportion of vertebrate to invertebrate foods varies seasonally: In winter fewer rodents are eaten, and the relative importance of invertebrate prey increases. Plant material is represented infrequently, and its proportion was not found to change significantly with the seasons. Jaksić *et al.* (1980) collected and analyzed 278 scats from central Chile. These were collected in all seasons except winter. These scats indicated that rodents dominate the diet. Chillas also consume a small number of birds, birds' eggs, snakes, and berries. The importance of plant materials, primarily berries, increases in the autumn. Rabbits are only rarely eaten. Jaksić *et al.* (1980) characterized chillas as opportunistic predators that hunt in patches of open vegetation.

Jaksić *et al.* (1983) presented data on the diet and trophic relations of chillas and culpeos (*P. culpaeus*) in both continental and insular parts of the Magallanes region of Chile (which lies along the Strait of Magellan). In this study they analyzed both scats and stomach contents. Here chillas prey primarily on mammals, and secondarily on birds, reptiles, invertebrates, and various plant material. In this region chillas are scavengers, and carrion makes up roughly one third of their intake. Diet studies indicate that sheep are not a major diet component (Duran and Cattan, 1985), although traditionally human control efforts have been based on the premise that sheep are an important prey of chillas.

ACTIVITY

The activity patterns of chillas are unclear. The best available evidence indicates that they are crepuscular (Jaksić *et al.*, 1980).

REPRODUCTION

No information on reproduction is available, though Nowak and Paradiso (1983) provided general comments on reproduction in the genus *Pseudalopex*.

SOCIAL ORGANIZATION AND BEHAVIOR

No information on social organization is available, although Nowak and Paradiso (1983) provided some general comments on social organization in the genus *Pseudalopex*. In southern Chile a field study estimated population density at 1 fox per 43 ha, with a total population of 65,800 foxes in 28,300 km² (Duran and Cattan, 1985).

Pseudalopex gymnocercus: Pampas Fox

One of the five extant species of foxlike canids in South America, *P. gymnocercus* is known by a variety of common names, among them pampas fox, pampas gray fox, Azara's fox, and Paraguayan fox (Crespo, 1975; Langguth, 1975b; Kleiman, 1967; Stains, 1975; Clutton-Brock *et al.*, 1976). This is one of the more common canid species in the central and eastern regions of Argentina, although in some areas its numbers have been considerably decreased due to intense hunting by humans, a diminishing food supply, and a decrease in refugia brought about by habitat destruction. Pampas foxes have been heavily hunted by man, primarily for their fur and also in retribution for preying on poultry. Their top speed of 58 kph allows humans on horseback to catch up to these foxes (Langguth, 1975b).

DISTRIBUTION AND HABITAT

Pampas foxes are found in east-central South America from southeastern Brazil to Paraguay, Uruguay, and northeastern Argentina as far south as the Rio Negro (Crespo, 1975; Langguth, 1975b; Cabrera, 1931; Ewer, 1973; Bueler, 1973). Due to intense land exploitation in east and central Argentina, pampas foxes have disappeared from most of the Province of Buenos Aires, south of Santa Fé, Córdoba, Entre Rios, and Uruguay (Crespo, 1975). They are found in a range of lowland habitats from the foothills of the Andes eastward to the Atlantic coast, on the pampas and humid grasslands of Brazil, in hilly regions, in deserts, in open forests, and in dry spiny thickets in the western part of their range in Argentina (Crespo, 1975; Langguth, 1975b; Cabrera, 1931; Ewer, 1973; Bueler, 1973). Langguth (1975b) stated that they are found as high as 4,000 m, though Crespo (1975) recorded that in Argentina they are primarily inhabitants of prairie environments from sea level up to only about 1,000 m. Pampas foxes dig their own dens or use burrows abandoned by armadillos or viscachas. They may also shelter in tree roots or rocky areas (Langguth, 1975b).

PHYSICAL CHARACTERISTICS

Pampas foxes are intermediate in size in relation to the other three species in the genus *Pseudalopex*. Langguth (1975b) stated that the head-plus-body length is 62 cm, while Bueler (1973) gives a range of 78.7–81.3 cm. Tail length is 33–35.6 cm (Bueler, 1973; Langguth, 1975b). Weight ranges from 4.8 to 6.5 kg according to Langguth (1975b), while Crespo (1975) gives a slightly lower average weight of 4.4 kg. The coat is heavy with dense fur and is thick enough to be used commercially in garments. Pelage is predominantly gray on the back and sides, sprinkled with black, and there may be a medial black stripe across the back. The underparts are pale to whitish. The legs are rusty red to yellowish-red and there may be a brown spot in the hollow of the leg. The paws are white to yellowish (Langguth, 1975b; Bueler, 1973; Nowak and Paradiso, 1983). The head, neck, and ears are rufous, and the intermaxillary region is dark gray to black. The dark muzzle distinguishes pampas foxes from culpeos (*P. culpaeus*); the throat is whitish. Long and bushy, the tail is tipped in black, and there is a black caudal gland spot (Langguth, 1975b; Kleiman, 1967). The skull resembles that of the red fox (*Vulpes vulpes*) and is similar to that of the culpeo (*P. culpaeus*). It differs from the latter only in having a slightly shorter, wider rostrum (Clutton-Brock *et al.*, 1976; Langguth, 1975b).

TAXONOMY

The generic status of this species is unclear. It has been placed in the genus *Canis* (Langguth, 1975b; van Gelder, 1978) as well as in the genus *Dusicyon* (Clutton-Brock *et al.*, 1976; Crespo, 1975; Cabrera, 1957; Osgood, 1934; Stains, 1975). However, Berta's (1987) recent analysis justifies retention in the genus *Pseudalopex*, at least for the time being. See Brum-Zorilla and Langguth (1980) for a discussion of the karyotype of *P. gymnocercus* (2n = 74), along with other members of the Canidae.

DIET

Pampas foxes are omnivorous. An analysis of their stomach contents in Argentina (n = 230) showed the following dietary composition: 75% of all food eaten was of animal origin; the remaining 25% was plant material, mostly fruit from trees and shrubs. The importance of plant material increased in the autumn. Approximately 14% of the food of animal origin was from domestic animals, primarily sheep, while the remaining 86% was from wild animals. Of these, 54% were mammals, primarily European hares (*Lepus europaeus*) and small rodents (pampas foxes play an important role in controlling wild rodent populations). The remainder was 31% birds and 10% insects. Altogether some 35 items were eaten. Pampas foxes also consume frogs, lizards, fish, sugar cane from cane plantations, and carrion (Langguth, 1975b). Pampas foxes allegedly endanger newborn lambs, and for this reason they are poisoned and hunted (Langguth, 1975b).

ACTIVITY

Usually nocturnal, pampas foxes are also active diurnally, particularly in regions where human activity is infrequent (Bueler, 1973; Nowak and Paradiso, 1983).

REPRODUCTION

Females are annually monestrous. They mate in August to October and bear their young in October or November (towards the end of spring in South America). After a gestation of roughly 58 days, one to eight pups are born; the usual litter size is three to five. The young begin accompanying their parents on hunting forays by December or January. In the wild, life expectancy is a few years. A captive lived for 13 years and 8 months (M. Jones, personal communication, cited in Nowak and Paradiso, 1983).

SOCIAL ORGANIZATION AND BEHAVIOR

These foxes seem to be solitary or semi-solitary, hunting singly or in pairs (Cabrera, 1940, cited in Kleiman, 1967). Langguth (1975b) stated that they are usually solitary and meet only during the mating season. Bueler (1973) recorded the same information, adding that occasionally two are found together during the summer. Nowak and Paradiso (1983, p. 944) commented that the male helps to provide food for the family (whether these authors refer to *P. gymnocercus*, *P. culpaeus*, or both is unclear). Based on behavioral observations of a single captive, Kleiman (1967, p. 371) remarked that, in terms of behavioral patterns, *P. griseus* has the most in common with jackals. It "seems as unrelated to the fox-like members of the family Canidae as the jackals are." Kleiman (1967) observed a single captive using a tail posture to express dominance that was identical to that used by coyotes (*Canis latrans*), wolves (*Canis lupus*), and jackals (*Canis spp.*), in which the tail is held in a straight line, parallel to the line of the back. Pampas foxes are not at all wary of humans. They sometimes exhibit a peculiar behavior of remaining completely motionless on the approach of humans.

Pseudalopex sechurae: Sechura Fox

The Sechura fox is found only in a small region on the northwest coast of South America. It is comparable in size and habitat preference to the kit fox (*Vulpes macrotis*) from North America. Common names include Sechuran fox, Sechura desert fox, and Peruvian desert fox (Birdseye, 1956; Stains, 1975; Clutton-Brock *et al.*, 1976; Asa and Wallace, 1990; Huey, 1969). As is also true for all members of its genus, little is known of its life history or behavior.

DISTRIBUTION AND HABITAT

Sechura foxes exist in the coastal zones of northwestern Peru and southwestern Ecuador (Huey, 1969; Birdseye, 1956; Cabrera, 1931; Ewer, 1973). These are desert animals that live in arid habitats, including sandy deserts and shifting dunes with low plant cover (Cabrera, 1931; Huey, 1969). They inhabit the Sechura Desert of northwestern Peru, a region described by Huey (1969, p. 1089) as "among the most desolate, arid areas in the Western Hemisphere." Birdseye (1956, p. 284) reported that they abounded along the Peruvian coast approximately 200 km north of Lima, where they inhabited rich cultivated areas, foothills, and beaches, as well as "nearly lifeless desert." In 1969 Huey reported tracks, droppings, and nocturnal sightings as common in the Sechura Desert. Current population levels are unknown.

PHYSICAL CHARACTERISTICS

This is the smallest species in the genus *Pseudalopex*, even smaller than the hoary fox (*Pseudalopex vetulus*). This small size is probably an adaptation to the demands and restrictions of desert life. Average weight is 2.2 kg (Huey, 1969). The pelage is similar to that of the hoary fox (*Pseudalopex vetulus*) and the chilla (*Pseudalopex griseus*). It is light in color with pale agouti guard hairs and fawn underfur. The underbody is cream to fawn. The muzzle is short, and the skull and carnassials small. The canines have been described as "fox-like" (Clutton-Brock *et al.*, 1976; Bueler, 1973). In terms of overall physical characteristics, the Sechura fox is most similar to the hoary fox (*Pseudalopex vetulus*) (Clutton-Brock *et al.*, 1976). Berta (1987) suggested that these two species represent a distinct lineage within *Pseudalopex*.

TAXONOMY

P. sechurae has often been placed in the genus *Dusicyon* (Clutton-Brock *et al.*, 1976; Stains, 1975; Birdseye, 1956; Huey, 1969) as well as in the genus *Canis* (van Gelder, 1978; Langguth, 1975b). Recent analysis, however, indicates that it should be regarded as one of the five extant species in the genus *Pseudalopex* Burmeister, 1856 (Berta, 1987). It is monotypic (Stains, 1975).

DIET

Sechura foxes are omnivorous. In some regions their diet is restricted by the availability of prey. Individuals inhabiting the Sechura Desert proper have the least varied food intake and are principally herbivorous during the winter (Asa and Wallace, 1990; Huey, 1969). At this time of year, they subsist primarily on seed pods of perennial shrubs, supplemented with tenebrionid beetles and, very rarely, a small vertebrate such as a lizard or bird. This seasonal herbivory (up to 99% by weight in the dry season in the Sechura Desert) is unique among the Canidae. After the rainy period of El Niño, grasshoppers and mice became important dietary components (Asa and Wallace, 1990). In coastal areas the diet is more varied, although seeds are still the only items in 63% of the scats analyzed. Sechura foxes in coastal areas also eat gulls, finches, various sea birds and their eggs, mice, lizards, snakes, crabs, beetles, and fish. Some of this is undoubtedly consumed as carrion (Huey, 1969; Birdseye, 1956; Asa and Wallace, 1990). A variety of cultivated fruits are eaten where available, including bananas, guavas, papayas, grapes, and mangoes (Birdseye, 1956). The captive kept by Birdseye (1956) was omnivorous, consuming fish, rodents, meat, eggs, insects, bread, and bananas. It would kill and consume poultry. Sechura foxes are able to survive without access to standing water, perhaps by licking surface condensation on foggy mornings (Asa and Wallace, 1990).

ACTIVITY

Sechura foxes are largely nocturnal (Birdseye, 1956; Huey, 1969). Four free-ranging foxes in the Sechura Desert of northwestern Peru began to forage before sunset and remained active throughout the night. During the day, the foxes remained in their dens, occasionally emerging to forage (Asa and Wallace, 1990). Nocturnal activity patterns were not influenced by lunar phases (Asa and Wallace, 1990).

REPRODUCTION

Almost nothing is known about reproduction in these foxes. Birdseye (1956) stated that all three he knew of were born in October or November.

SOCIAL ORGANIZATION AND BEHAVIOR

Almost nothing is known about the social organization of free-ranging Sechura foxes. In the Sechura Desert, four radio-collared foxes occupied two separate home ranges. A single male inhabited one range, and an adult female with two subadults occupied the other (Asa and Wallace, 1990).

Birdseye (1956), who kept a wild-caught female from the age of about 10 days, included some casual observations on her behavior. She was kept entirely without contact with conspecifics. This female became quite tame, although she always remained nervous and was "much disturbed" by strangers. She actively sought the company of her keepers and seemed to enjoy being scratched. Before becoming full grown she was inquisitive and quite playful. By the age of 18 months she played less frequently and usually only by herself. She urine-marked frequently and exhibited site-specific defecation behavior.

A variety of vocalizations emitted by a single human-reared captive are described in Birdseye (1956).

Pseudalopex vetulus: Hoary Fox

This is a small South American canid quite similar in physical appearance to the four other species in the genus *Pseudalopex*. As is also the case for the other *Pseudalopex* species, little is known about the life history and behavior of *Pseudalopex vetulus*. Coimbra Filho (1966) remarked on this fact some 20 years ago, and his statement still holds true. Although these foxes are not particularly rare, no one has yet completed a field study of any aspect of their life. Common names include hoary fox, field fox, and small-toothed dog.

DISTRIBUTION AND HABITAT

These foxes occur in central, south-central, and east-central Brazil in the states of Mato Grosso, Goiás, Minas Gerais, and São Paulo (Langguth, 1975b; Clutton-Brock *et al.*, 1976; Coimbra Filho, 1966; Ewer, 1973; Bueler, 1973; Stains, 1975; Nowak and Paradiso, 1983). Hoary foxes "may represent the typical grassland canid in the open country of central Brazil" (Langguth, 1975b, p. 200). Preferred habitat is open country, such as grassy savanna or the *campos cerrados* (savannas with scattered trees). The foxes may take shelter in the burrows of armadillos (Langguth, 1975b; Nowak and Paradiso, 1983).

PHYSICAL CHARACTERISTICS

Head-plus-body length ranges from 58 to 64 cm, tail length from 28 to

32 cm, and weight from 3.6 to 4.1 kg. The pelage is short with an over-all gray tone, though the coats of some individuals may be lighter than the usual. The underparts are light, fawn or cream in color. The head is gray with white on the throat, and there is black on the point of the jaw. Ears are tawny with black tips. The snout is shorter than in red foxes (*Vulpes vulpes*). The legs are gray, tending to yellow, red, or tawny on their lateral edges. The tail continues the color of the back, and there is a dark stripe along its dorsal line. The tail tip is black, and there is a black caudal gland spot (Langguth, 1975b; Clutton-Brock *et al.*, 1976; Bueler, 1973; Nowak and Paradiso, 1983).

The muzzle is relatively short, and the skull and teeth are small. The molars are broader than in red foxes. The carnassials are reduced in proportion to the molars (Berta, 1987), suggesting a less than strictly carnivorous diet and perhaps a reliance on insects (Langguth, 1975b). The canines are sharply pointed and foxlike (Langguth, 1975b; Clutton-Brock *et al.*, 1976; Stains, 1975; Bueler, 1973; Nowak and Paradiso, 1983). As in *P. sechurae*, the rostrum is proportionately shorter than those of the other *Pseudalopex* species (Berta, 1987).

TAXONOMY

Often placed in the genus *Dusicyon* (Stains, 1975; Clutton-Brock *et al.*, 1976; Cabrera, 1957), the hoary fox has also been included in the genus *Canis* (van Gelder, 1978). Langguth (1975b) grants the hoary fox full generic rank, as *Lycalopex*. In short, its taxonomic status is disputed. At least for the short term, Berta's (1987) taxonomic analysis justifies its retention in the genus *Pseudalopex* Burmeister, 1856. The species seems to be monotypic (Stains, 1975). See Wurster and Benirschke (1968) for karyotypic information (2n = 74: NF = 76).

DIET

No stomach content or scat analyses have been published for these foxes. They feed on small rodents, birds, and insects, particularly grasshoppers (Santos, 1945, and Lund 1950, cited in Langguth, 1975b). A captive caught rats, mice, and insects and would eat any sort of animal food, rejecting all foods of vegetable origin (Lund, 1950, cited in Langguth, 1975b). A 3.8 kg captive at the Rio de Janeiro Zoo subsisted on meat, raw eggs, and bananas (Coimbra Filho, 1966). Coimbra Filho

(1966) stated that hoary foxes eat a higher proportion of foods of animal origin than do crab-eating foxes (*Cerdocyon thous*). Hoary foxes are persecuted by humans because they are thought to kill poultry (Langguth, 1975b; Coimbra Filho, 1966).

ACTIVITY

These foxes are active during the day and early evening (Langguth, 1975b).

REPRODUCTION

The usual range in litter size is two to four, born in September (springtime). The female dens in a deserted armadillo burrow or similar deep cover (Langguth, 1975b; Coimbra Filho, 1966; Bueler, 1973). Coimbra Filho (1966) reported on a captive pair who bred at the Rio de Janeiro Zoo, producing four pups, which were soon abandoned, probably because their enclosure was inadequate.

SOCIAL ORGANIZATION AND BEHAVIOR

Practically nothing is known of the social organization or behavior of *Pseudalopex* or of its relationship with other South American members of the family Canidae. No field studies of behavior or social organization have been done. Reporting on the life of a captive pair housed at the Rio de Janeiro Zoo, Coimbra Filho (1966) stated that they generally lived harmoniously. On one occasion the male attacked the female, and the animals were then separated for a short period and subsequently reunited. No other disputes followed. This same pair reproduced successfully but abandoned their young soon after birth for unknown reasons.

Bush Dogs *(Speothos venaticus)*
Credit: Tom McHugh/National Audubon Society
Collection/Photo Researchers, Inc.

CHAPTER 13
Genus *Speothos*

Speothos venaticus: Bush Dog

Bush dogs are peculiar animals, physically unlike almost all other members of the family Canidae. The only canid to which bush dogs appear similar is the small-eared dog, *Atelocynus microtis*. Bush dogs are rarely seen in the wild, and then only fleetingly. Almost all behavioral observations have been made of captive individuals, and virtually nothing is known of the behavior and life history of free-ranging animals (Deutsch, 1983; Porton, 1983; Brady, 1982; Collier and Emerson, 1973; Langguth, 1975b). A considerable number of captivity studies have recently been undertaken (Brady, 1981, 1982; Porton, 1983; Porton *et al.*, 1987; Biben, 1982a,b, 1983; Drüwa, 1976; Kleiman, 1972), and there are several accounts of zoo-kept animals as well (Jantschke, 1973; Collier and Emerson, 1973; Kitchener, 1971). Classified as vulnerable by the IUCN, bush dogs seem to disappear as human activities encroach on their habitat. They seem to be particularly affected by forest destruction.

DISTRIBUTION AND HABITAT

Bush dogs are found in central and north-central South America, from the northeastern edge of Argentina north through Paraguay and the eastern regions of Bolivia, Peru, and Colombia, north into Panama, and east into Venezuela, the Guianas and most of northern and central Brazil (Berta, 1984; Clutton-Brock *et al.*, 1976; Langguth, 1975b; Nowak and Paradiso, 1983; Stains, 1975). They inhabit tropical rainforests and are found along forest borders and in wet savannas. They seem to prefer habitats close to water and have been characterized as semi-aquatic (Biben, 1982a; Langguth, 1975b; Bates, 1944; Clutton-Brock *et al.*, 1976). Bates (1944) remarked on his captive's remarkable swimming ability.

PHYSICAL CHARACTERISTICS

Bush dogs have a stocky, compact body with a broad muzzle, small ears, and short legs and tail. Head-plus-body length ranges from 57 to 75 cm, with a height at shoulder of about 30 cm and a tail length of 12–15 cm. Weight ranges from 5 to 7 kg (Langguth, 1975b; Hershkovitz, 1957; Clutton-Brock et al., 1976). The pelage is a sleek, uniform dark brown, with no conspicuous facial or body markings and, unlike most canids, no countershading. The neck, hackles, and ears are lighter in color and are often described as ocherous. There may be a lighter patch on the throat, and pale patches or bands may be present on the body (Hershkovitz, 1957; Clutton-Brock et al., 1976). Ewer (1973; citing Langguth, 1969) stated that a caudal gland is present, although Kleiman (1972) recorded that no caudal glands were evident in the specimens she examined. The tail is well furred but not bushy (Hershkovitz, 1957).

The dentition of bush dogs seems to be specialized for a heavily carnivorous diet (Kleiman, 1972). The mandible is short and robust, as are the canines (Stains, 1975). The upper and lower second molars are reduced; the upper second molar may be absent. The lower third molar is always absent, a peculiarity shared by the dholes (Cuon alpinus), giving a dental formula of incisors 3/3, canines 1/1, premolars 4/4, molars 1/1 or 2 = 38 or 40 (Clutton-Brock et al., 1976; Stains, 1975; Hershkovitz, 1957). The usual canid dental formula is incisors 3/3, canines 1/1, premolars 4/4, molars 2/3 = 42.

TAXONOMY

Speothos is universally recognized as the monospecific genus Speothos Lund, 1839. Past classification schemes placed Speothos within the subfamily Simocyoninae which contained two other monospecific genera as well, Lycaon (African wild dogs) and Cuon (dholes) (Simpson, 1945; Stains, 1975). This was a thoroughly heterogeneous assemblage by any stretch of the imagination, and the classification was based only on several dental peculiarities within the group. The original diagnostic characteristic of Simocyoninae had been "the development of the talonid of the lower carnassial as a single cusp or ridge" (Clutton-Brock et al., 1976, p. 178): In all other canids the talonid has two cusps. Whatever the validity of postulating a common origin of these three genera, they have diverged greatly and ought to be recognized as isolated and distinct (Clutton-Brock *et al.*, 1976). In addition, there is an overall lack of behavioral similarity between *Lycaon* and *Speothos*, and on the basis of behavior alone, they are "probably as widely separated from each other as they are from each member of the Caninae" (Kleiman, 1967, p. 371). In sum, the subfamily Simocyoninae should be discarded as a taxonomic entity. Recent reclassifications have recognized two subfamilies within the family Canidae: the subfamily Otocyoninae, represented by the bat-eared fox, and the subfamily Caninae, containing all other genera (Anderson and Jones, 1984). Berta (1984, 1987) discussed the taxonomic position of *Speothos* in detail, suggesting *Atelocynus microtis* (small-eared dogs) as the closest relative. See Wayne and O'Brien (1987) for an alternative assessment, based on electrophoretic data from blood and tissue samples. They suggested that the short limbs and reduction in number of post-carnassial teeth might be a consequence of dwarfism rather than an adaptation to particular environmental constraints. Three subspecies are recognized (Stains, 1975).

DIET

Reports on dietary composition are largely anecdotal or unattributed, and no analyses of stomach contents or scats have been published for free-ranging bush dogs. They are reported to feed on a wide variety of items, including aquatic invertebrates and vertebrates, birds, various small terrestrial animals, and larger South American rodents, such as capybara and paca (Deutsch, 1983; Kleiman, 1972; Langguth, 1975b). They may capture prey as large as deer, but the primary food source seems to be large rodents (Deutsch, 1983; Hershkovitz, 1957; Langguth, 1975b; Kleiman, 1972). Bush dogs probably hunt in packs, which would seem to be a necessary strategy given the relatively large size of their preferred prey. There is only one field observation of their hunting techniques (Deutsch, 1983; Biben, 1982b).

Captives attack and consume birds and rodents (Kitchener, 1971), and Bates (1944) remarked that anything other than dogs or men were regarded as potential food by a bush dog raised by humans. Collier and Emerson (1973) also stated that captives treated any items smaller than their own size as potential prey.

ACTIVITY

Bush dogs seen to be primarily diurnal (Kleiman, 1972; Deutsch, 1983; Nowak and Paradiso, 1983), although Langguth (1975b) stated that they are active both diurnally and nocturnally.

REPRODUCTION

All information on reproduction in bush dogs has been obtained from observations of captives. Five captive females in Virginia had no fixed reproductive season. Their estrous periods occurred in every month of the year, and their pups were born throughout the year (Porton *et al.*, 1987). Bush dogs live only in tropical habitats, and therefore their reproduction may be released from the seasonal constraints imposed in temperate climates. "In addition to climate, seasonal variation in food availability most likely influences the reproductive timing of tropical species. Bush dogs probably prey on medium-sized mammals. . . . These species are typically stable in numbers and density. . . . A relatively stable prey base may have allowed year-round breeding by bush dogs and thereby permitted the timing of reproduction to be influenced by social factors" (Porton *et al.*, 1987, p. 870). This reproductive aseasonality may also occur in African wild dogs (*Lycaon pictus*), crab-eating foxes (*Cerdocyon thous*), and, in some regions, bat-eared foxes (*Otocyon megalotis*) and golden jackals (*Canis aureus*) (Porton *et al.*, 1987; Malcolm, 1986). Not all tropical canids have this high degree of reproductive flexibility, however; maned wolves (*Chrysocyon brachyurus*), for example, do not.

Two young primiparous females had multiple estrous periods before conception (Porton *et al.*, 1987). Females kept alone or with members of their natal group exhibited reproductive suppression. Removal from this situation and placement with a male resulted in initiation of estrus (in 9.5-month or older females). Thus a clear social influence on reproduction exists (Porton *et al.*, 1987). Females in estrus assume a lordosis posture, which seems to be unique among the Canidae (Kleiman, 1968). Reports give gestation periods from 65 to over 80 days (Crandall, 1964, cited in Kleiman, 1972; Kitchener, 1971). A likely average is 67 days (Porton *et al.*, 1987). Litters range from one to six pups (Collier and Emerson, 1973; Jantschke, 1973; Langguth, 1975b; Husson, 1978; also cited in Nowak and Paradiso, 1983). Loss or removal of young pups results in a significantly reduced time until onset of the next estrous period (Porton *et al.*, 1987).

Few descriptions of parental care behaviors are available, since captive-born pups are often removed from the den by human caretakers soon after birth. Porton (1983, p. 1067) stated that males exhibit "more direct parental care than has so far been described for any other canid (and most mammals)." With the obvious exception of lactation, males and females show very similar care-giving behavior patterns throughout the pup-rearing period. The male's presence may be a requisite for appropriate maternal care-giving behaviors on the part of the female (112th Annual Report to the Frankfurt Zoological Garden, 1970, cited in Kleiman, 1972). Indeed, the father's presence may be required for successful pup rearing (Jantschke, 1973). Females may compete for the high level of parental investment exhibited by males (Porton, 1983). The social organization of bush dogs is characterized by the intense female-female competition found only in other obligately monogamous mammals.

A captive bush dog lived for over 10 years (M. Jones, personal communication, cited in Nowak and Paradiso, 1983).

SOCIAL ORGANIZATION AND BEHAVIOR

Bush dogs appear to be among the most highly social members of the Canidae family. Sightings of free-ranging groups of 7–12 animals have been made (Deutsch, 1983; Defler, 1986), and there has been one sighting of a group of hundreds of individuals (Husson, 1978). Brady (1982) stated that bush dogs live in extended families, which perhaps evolve through the continued interaction of parents and offspring after the weaning period (Kleiman and Eisenberg, 1973). The evidence for this assertion however, is based almost wholly on observations of captives' behavior. Biben (1982a) suggested that the fundamental social unit of bush dogs is either the pack, the mated pair, or the mated pair accompanied by subadult offspring. Individuals may also hunt alone (Deutsch, 1983).

In captivity, bush dogs maintain strong pair bonds, exhibit shared parental care of offspring, and show high rates of affiliative behaviors, such as sleeping in contact, hunting cooperatively, and having prominent greeting and submission ceremonies (Kleiman, 1982; Biben,

1982b; Porton, 1983; Drüwa, 1976). Adult captives can be kept in same- or mixed-gender groups, although unfamiliar individuals may fight until a dominant/subordinate relationship develops (Kleiman, 1972). Mutual submission ceremonies are prominent in mated pairs and family groups, where they probably function to promote group cohesion (Brady, 1982).

Vocal signals and close-range appeasement behaviors are also prominent among bush dogs (Brady, 1982; Biben, 1982b). Levels of intraspecific aggression are low, and overt competition for food, when it occurs, is not aggressive. Among bush dog pups, submissive behaviors and vocalizing are frequent, while competition for access to play objects is minimal (Biben, 1982b). One typical canid behavior conspicuous by its absence in *Speothos* is the stereotyped play bow (Biben, 1982b).

Food sharing among bush dogs is highly developed (Biben, 1982a). Captive pups never defend or fight over food or potential prey. Pups and parents do shove each other to gain access to food, but they never threaten each other or retaliate when jostled (Biben, 1982a). This lack of aggression (or this high degree of tolerance) with respect to feeding does not appear to be learned by bush dog pups, but rather occurs without prior feeding- or nonfeeding-related experiences (Biben, 1982a).

Urine-marking behaviors are prominent. Bush dogs mark themselves and each other, as well as objects in their environment (Brady, 1982). To urine-mark, females assume a handstand position. Their stance may function as a visual signal in sexual recognition (Porton, 1983). African wild dogs (*Lycaon pictus*) may also handstand when urine-marking environmental objects, but they do so only briefly (Buitron, 1977) and do not back up to environmental objects and place their rear paws against them, as female bush dogs will do. Urine marking of self and pair mate seems to function in the formation and maintenance of pair bonds (Kleiman, 1972; Brady, 1982; Porton, 1983). Urine-marking behaviors are also prominent during agonistic encounters (Biben, 1982c). In captivity, bush dogs deposit feces throughout their enclosures, as opposed to favoring distinct sites (latrines or middens) (Kleiman, 1972).

Bates (1944) reported that a hand-reared female adapted readily to captivity. She interacted frequently and eagerly with familiar humans and displayed behavior patterns strikingly similar to those of domestic dogs. She also had a remarkable swimming ability.

Partially as a result of quantitative comparative studies of bush dogs, maned wolves (*Chrysocyon brachyurus*), and crab-eating foxes (*Cerdocyon thous*), the conventional idea that the more social species of canids have more complex, less stereotyped behavioral repertoires has been called into question (Brady, 1981, 1982; Biben, 1983; Kleiman, 1967; Fox, 1975). Quantitative comparative data on bush dogs (highly social), crab-eating foxes (moderately social), and maned wolves (among the least social of the Canidae) show that the relative complexity of the behavioral repertoires of each species does not correspond to their degree of sociality (Brady, 1982, pp. 7–8, 104–120; Biben, 1983, p. 824). However, major interspecific differences in the relative frequencies of particular behavior patterns are evident. Specifically, although bush dogs are more social than maned wolves, they have a more limited behavioral repertoire and more rigidly determinate intraspecific interaction sequences than both maned wolves and crab-eating foxes (Biben, 1983, p. 823).

Gray Fox (*Urocyon cinereoargenteus*)
Credit: R. Austing/National Audubon Society
Collection/Photo Researchers, Inc.

CHAPTER 14
Genus *Urocyon*

Urocyon cinereoargenteus: Gray Fox

There are two species in the genus *Urocyon*: the gray fox (*U. cinereoargenteus*) and the island gray fox (*U. littoralis*), an insular species found only on islands off the southern coast of California. Gray foxes are common, inconspicuous, medium-sized canids. They are highly adaptable, and their habits are similar to those of red foxes (*Vulpes vulpes*). Their fur, though not of premium quality, is widely sold commercially, which inevitably creates pressure on the species from hunting and trapping activities. Trapp and Hallberg (1975) and Fritzell (1987) surveyed the ecology, life history, and social behavior of this species. See Fritzell (1987) for an extensive bibliography.

DISTRIBUTION AND HABITAT

Regions along the U.S.–Canadian border, both to the east and west of the Great Lakes, are the northernmost limit of distribution. From here, gray foxes occur southward through all of North America except for the northern Rocky Mountain region, the northern reaches of the Great Basin, and Washington state. The range extends south through Central America and into the northwestern corner of South America in Venezuela and Colombia (Fritzell and Haroldson, 1982; Banfield, 1974; Trapp and Hallberg, 1975).

As might be anticipated from the size of their range, which includes two continents, gray foxes are found in many and varied habitats. In the eastern United States, habitats include deciduous woodland, forest borders, and, in Florida, scrubby woodland, old fields, citrus groves, and railroad rights-of-way (Wasmer, 1984; Carr, 1945). In Texas, gray foxes occupy post-oak woodland, pinon-juniper woodland, and wooded sections of short grass plains (Fritzell and Haroldson, 1982; Trapp and Hallberg, 1975). In western North America, they inhabit brushy regions, woodland, rugged or broken terrain, chaparral, pinon-juniper woodland, washes, mountainsides, and brushy meadows (Trapp and Hallberg, 1975; Fritzell and Haroldson, 1982). In South America gray foxes are found in dry, open country, and in Venezuela they occur in scrub lands (Hershkovitz, 1957; Tate, 1931). Throughout their range gray foxes may be found in association with cultivated areas and city outskirts. Dens are used year-round but are most important during whelping season. They are often located in wooded, brushy, or similarly sheltered areas, such as rock crevices or outcrops, and brush or weed piles, and in hollow logs. Hollow trees are used, and tree dens have been found up to 9.1 m above the ground. Underground burrows are also used. These are either the abandoned burrows of other animals, or ones dug by the foxes. Gray foxes are more likely to use underground burrows in the northern part of their range (Fritzell and Haroldson, 1982; Nicholson et al., 1985; Banfield, 1974; Tate, 1931). Unlike all other canids [with the possible exception of corsac foxes (Vulpes corsac); Langguth, 1975a], gray foxes readily climb trees and are able to scale limbless trunks and jump from one branch to another. Gray foxes will often seek refuge in trees when pursued and may rest or forage there as well (Carr, 1945; Fritzell and Haroldson, 1982; Taylor, 1943; Banfield, 1974).

PHYSICAL CHARACTERISTICS

Head-plus-body length ranges from 48 to 73 cm. Tail length is 27–44 cm. Weight ranges from 2.5 to 7 kg, and males are slightly larger than females. Banfield (1974) gave the average weight of males as 4.1 kg (range 3.4–5.9 kg) and that of females as 3.9 kg (range 3.4–5.1 kg). Juveniles reach adult weight by the age of six months. The legs of gray foxes are proportionately shorter than those of red foxes; therefore gray foxes appear smaller although they actually weigh about the same (Fritzell, 1987). Overall, the fur is grizzled gray and relatively coarse. Guard hairs are banded in gray, black, and white. The underfur is buff to gray. Underparts are pale whitish to white. There is a dark stripe extending along the dorsal midline and continuing down the tail. Portions of the sides may be cinnamon-rufous in color (Fritzell and Haroldson, 1982; Banfield, 1974; Clutton-Brock et al., 1976). The throat and jaws are white, and the sides of the neck, cheeks, and head are ocher to rusty.

The chin is gray or brown, and there is a black muzzle patch below each eye and on the lower jaw. The ears are ocher or tawny (Banfield, 1974; Hall and Kelson, 1959; Fritzell and Haroldson, 1982; Clutton-Brock *et al.*, 1976). The tail is long and well furred, rust in color on its underside, with a median dorsal stripe of black coarse hair which continues the black dorsal stripe of the back. The caudal gland is large, apparently extending one third to one half the length of the tail. This is the largest caudal gland of any canid (Hildebrand, 1952, cited in Clutton-Brock *et al.*, 1976); it is covered by a ridge of stiff black guard hairs.

Despite many similarities to red foxes and placement in the genus *Vulpes* by Clutton-Brock *et al.* (1976), the skull and dentition of the gray fox are quite different from those of *Vulpes vulpes* (Banfield, 1974; Fritzell, 1987). The teeth are well developed, although the canines are not as long as is usual for vulpines. The carnassials and molars are "fox-like" (Clutton-Brock *et al.*, 1976). The dentition conforms to the usual canid pattern of incisors 3/3, canines 1/1, premolars 4/4, molars 2/3 = 42. The forearm has greater rotational mobility than that of any other species of the Canidae; This may be related to gray foxes' unusual tree climbing ability (Ewer, 1973).

TAXONOMY

There are two species in the genus *Urocyon*, although whether the island gray fox (*U. littoralis*) deserves specific status is unclear (Fritzell and Haroldson, 1982; Stains, 1975). There are 15 or 16 subspecies (Fritzell and Haroldson, 1982; Stains, 1975; Hall and Kelson, 1959). Some authorities have included the gray fox within *Vulpes* as *Vulpes cinereoargenteus* (Macdonald, 1984; Clutton-Brock *et al.*, 1976). However, morphologically, karyotypically, and biochemically, enough significant differences exist to warrant separate generic status (Wayne and O'Brien, 1987). In North America, fossil representation of *U. cinereoargenteus* is extensive, beginning 1 million years ago. South American fossil representatives are unknown (Berta, 1987).

DIET

Generally, gray foxes are omnivorous, opportunistic feeders. Feeding patterns may vary considerably with season and locale. The diet includes vertebrates, insects, and plant material. Small mammals are generally most important, although during some parts of the year plant material may be the chief dietary component (Fritzell and Haroldson, 1982; Trapp and Hallberg, 1975; Banfield, 1974). In Zion National Park, Utah, gray foxes are herbivorous, insectivorous, and scavengers, more than carnivorous (Trapp and Hallberg, 1975). They eat mammals including lagomorphs, primarily cottontail rabbits; rodents, such as deer mice, harvest mice, cotton rats, squirrels, woodchucks, shrews, and pocket gophers; and opossums. They consume deer in the form of carrion. Predation on domestic stock seems to be unusual (Trapp and Hallberg, 1975; Banfield, 1974; Fritzell and Haroldson, 1982). Gray foxes eat pheasants, ducks, chickens (primarily as carrion), and a variety of passerine birds (Ewer, 1973; Trapp and Hallberg, 1975; Banfield, 1974). Insects may make up as much as 41%, by volume, of food eaten. Orthopterans, Coleopterans, and Lepidopterans are all consumed (Trapp and Hallberg, 1975; Banfield, 1974; Langguth, 1975a). In autumn plant foods may form up to 70% (by volume) of the diet (Fritzell and Haroldson, 1982). Gray foxes eat fruit (persimmons, apples, grapes, juniper, and prickly pear fruit), nuts (hickory and beech nuts, acorns, peanuts), grains, corn, and grasses (Ewer, 1973; Fritzell and Haroldson, 1982; Trapp and Hallberg 1975). Fish are eaten when available (Banfield, 1974). In the eastern and north-central United States the autumn diet (in decreasing percentage of representation) is composed of mammals, plants, birds, and carrion. In more southerly regions, during the winter, arthropods appear in the diet, although mammals still have the greatest volume. In Texas in springtime, arthropods, plants, and reptiles are important although less so than mammals (Trapp and Hallberg, 1975). Captives reared by humans are omnivorous, consuming milk, mice, birds, eggs, bananas, apples, and dog chow (Taylor, 1943).

ACTIVITY

Members of this species are nocturnal and/or crepuscular. During the day they remain concealed, occasionally moving from one resting place to another (Wasmer, 1984; Fritzell and Haroldson, 1982; Sunquist, 1989; Trapp and Hallberg, 1975; Banfield, 1974).

REPRODUCTION

Females are annually monestrous. Most reach sexual maturity in their first year. The breeding season varies geographically. In southern latitudes (Florida and Georgia) estrus occurs in January and February, while in northern regions breeding is later, as late as mid-May in New York (Fritzell and Haroldson, 1982; Fritzell, 1987; Trapp and Hallberg, 1975; Banfield, 1974). See Follmann (1978) for data on the male reproductive cycle. Gestation in captivity is about 59 days (Fritzell, 1987). Fifty-three days is the typical gestation period of red foxes (*Vulpes vulpes*), and 63 days the usual for species in the genus *Canis*. Parturition occurs anywhere from mid-March in the south to mid-June in the north. Litters range from one to ten pups, with the mean at about four (Fritzell and Haroldson, 1982; Fritzell, 1987; Nicholson *et al.*, 1985; Banfield, 1974; Wood, 1958). Monogamy seems to be the usual breeding strategy, with polygyny occurring occasionally as well (Banfield, 1974; Fritzell and Haroldson, 1982; Trapp and Hallberg, 1975). Males may assist in pup rearing (Banfield, 1974). Males of mated pairs forage separately from the females at night and do not return repeatedly to the den as the females do (Fritzell and Haroldson, 1982). Two females may rear their litters in the same den (Banfield, 1974), which may account for Nowak and Paradiso's (1983) somewhat inflated upper limit for litter size (10). By the time the pups are 3 months old they accompany their parents on foraging expeditions away from the den, and by the age of 4 months they forage independently. They remain in the parental home range until January or February of the following year. Pups reach adult size early in winter (Fritzell and Haroldson, 1982; Trapp and Hallberg, 1975; Nicholson *et al.*, 1985).

Gray foxes are preyed on by pumas, eagles, and coyotes. The pups are killed by large raptors (Fritzell and Haroldson, 1982; Trapp and Hallberg, 1975; Ewer, 1973). Man remains the most significant predator, killing approximately 370,000 in the 1979–80 season in the United States (Fritzell, 1987). As much as half the gray fox population may be "harvested" annually in Wisconsin (Fritzell, 1987). Maximum life span of a captive was 13 years and 8 months (Nowak and Paradiso, 1983).

SOCIAL ORGANIZATION AND BEHAVIOR

Mated pairs and their offspring form the essential social unit. Each family unit generally maintains a separate home range (Fritzell and Haroldson, 1982). Home range sizes vary considerably. This variance is at least partially a function of habitat quality and the distribution of resources in that habitat (Fuller, 1978). Gender and season also affect home range sizes of individuals (Fritzell and Haroldson, 1982). Males have larger ranges than females. Population density probably plays a role in the size of home ranges (Fuller, 1978). Fritzell and Haroldson's (1982) estimates of home range size, based on radiotelemetry studies, are 0.4–3.2 km². Other estimates vary from 30 ha to 2,755 ha (Fritzell, 1987). See Fritzell (1987) for a detailed summary of home range sizes and population densities in North American gray fox populations. Fuller's (1978) radiotelemetry study of the movements of four females in California gave a range of 30–40 ha. Wasmer's (1984) radiotelemetry study of an adult female tracked for 2 months in Florida gave a home range size of 2.69 km², consistent with other home range sizes in the southeastern United States. Sunquist (1989) found a mean home range size of 550 ha for three Florida gray foxes. Home range sizes increase in the late fall and winter. During whelping season they are at a maximum for males and much reduced for females (Nicholson *et al.*, 1985; Fritzell and Haroldson, 1982). Subadults disperse considerable distances, up to 84 km from the natal area (Sheldon, 1953). Banfield (1974) stated that tagged juveniles dispersed as far as 135 km. After the young have dispersed, the males tend to leave their mates (Banfield, 1974).

A number of estimates for population densities have been published. Fritzell and Haroldson (1982) estimated from 1.2 to 2.1 foxes per km². Nowak and Paradiso (1983) gave a range of 0.4–10 per km². Trapp and Hallberg (1975) gave a range of 1–27 foxes per 2.6 km².

The issue of territoriality remains unclear. Family groups appear to exist in separate areas in Florida and perhaps also in Zion National Park, Utah (Trapp and Hallberg, 1975). Scent marking probably plays a role in maintenance of these areas: Urine and feces are left in conspicuous locations or along travel routes, and feces are often deposited in groups (Fritzell and Haroldson, 1982; Trapp and Hallberg, 1975). Captives cover urine and feces of conspecific cage mates with their own scent marks, and pairs may sequentially scent-mark 10–15 times in the same spot (Fox, 1971a). Taylor (1943) gave a detailed account of five mixed-sex captives raised by humans. Caged until 15 months old, they were then allowed to roam freely. As young pups housed together they were "extremely quarrelsome" and consistently fought. They were also playful, and continued to be playful at the age of 10 months. By then their play was less aggressive. From the age of 6 months onward they slept close together. Touching noses and licking inside each other's ears were frequent behaviors. Taylor remarked on the conspicuous individual variation in temperament, noting that some were "gentle, tame and cooperative," while others were much less so. Fox (1970, 1971a), working with captives reared by humans, made a number of observations. He recorded (1971a) that gray foxes dominate red and arctic foxes in competitive interactions. Pair bonding is "more enduring" in gray than in red foxes. His studies of a single male and a single female, captive and reared by humans, included observations on behavioral development, behavior, and the development and use of facial expressions (Fox, 1971a, pp. 52–53). Fox and Cohen (1977) included observations of captives and remarked on the large amount of social grooming in this species.

Mewing, a vocalization emitted only by foxes among adult members of the Canidae, appears in contexts of greeting or submission, as part of contact-seeking behavior, and possibly as a sign of distress (Cohen and Fox, 1976; Tate, 1931). In the nonvulpine Canidae species, mewing does not occur after the neonatal period. As for all members of the Canidae, gray foxes growl in contexts of threat and defense (Marten, 1980; Cohen and Fox, 1976). Snarls function as high intensity agonistic vocalizations, on a continuum with, and grading into, growls. Barks occur in contexts of alarm and threat (Taylor, 1943; Carr, 1945; Tembrock, 1963a,b). Screams occur in contexts of defense, distress, greeting, and contact seeking. (Marten, 1980; Cohen and Fox, 1976). Coos are a long-range greeting or contact-seeking call (Marten, 1980; Cohen and Fox, 1976). This particular vocalization occurs only among the vulpine species of the Canidae.

Urocyon littoralis: Island Gray Fox

Island gray foxes have an extremely restricted insular distribution. The species is listed as threatened by the state of California, and is being considered for listing as endangered or threatened by the U.S. Fish and

Wildlife Service (Ginsberg and Macdonald, 1990). The natural history of the species is quite similar to that of gray foxes (*U. cinereoargenteus*).

DISTRIBUTION AND HABITAT

Island gray foxes occur on six out of the eight Channel Islands off southern California's coastline (San Clemente, Santa Catalina, Santa Rosa, Santa Cruz, San Miguel, and San Nicolas Islands). The largest of these islands, Santa Cruz, is only 38 km long by 3–13 km wide. Two of these islands are administered by the National Park Service, and two by the U.S. Navy.

Climate on the islands is maritime Mediterranean; see Laughrin (1977) for information on habitat and vegetation characteristics. Island gray foxes occur in a wide variety of habitats, including sand dunes, open forest, grassland, and coastal scrub. Dens are in rocks or brush, or in simple tunnels. Sometimes manmade structures are used. Some dens are reused in successive years, others are not.

PHYSICAL CHARACTERISTICS

Island gray foxes are approximately 20% smaller than mainland gray foxes. In other respects the two species appear quite similar. Head-plus-body length is 48–50 cm; tail length is 11–29 cm. Reported adult weights range from 2.1 to 2.8 kg (Ginsberg and Macdonald, 1990; Laughrin, 1977). Mean weight of 15 individuals from one study conducted in early spring was 1.6 kg (Laughrin, 1977). This is a lower average than is generally reported. Island gray foxes have two fewer tail vertebrae than gray foxes (Laughrin, 1977). Like those of mainland gray foxes, island foxes' toenails are long, which facilitates climbing.

TAXONOMY

This insular species is closely related to its mainland congener *U. cinereoargenteus*. There is some debate whether it deserves specific status (Fritzell and Haroldson, 1982; Clutton-Brock *et al.*, 1976; Stains, 1975). There are six subspecies, one from each island. See Gilbert *et al.* (1990) for a discussion of the genetic relatedness and variability of the subspecies, based on restriction fragment polymorphisms.

Island gray foxes dispersed from the mainland to the Channel Islands during the late Pleistocene when the northern islands were a single landmass and the overwater distance was on the order of 6.5 km. The southern populations may have been transported to the islands by colonizing Native Americans some 10,000 years ago (Gilbert *et al.*, 1990; Fritzell, 1987). Despite recent debate, it appears that the foxes on San Clemente have been present on the island since prehistoric times. They were not introduced by humans in 1875, although some period correspondence suggested that an additional pair of foxes from Santa Catalina Island may have been introduced at that time (Johnson, 1975).

DIET

Island gray foxes are opportunistic and omnivorous. Scat analysis shows a primary reliance on insects (primarily Coleopterans and Orthopterans) and seasonally available fruit (Laughrin, 1977). The foxes eat some birds. There are relatively few rodents on the islands, and their remains are correspondingly infrequent in scats (Laughrin, 1977). Laughrin (1977) suggested that island gray foxes' restricted diets may be a limiting factor in their size.

ACTIVITY

Island gray foxes are active at all times of the day and night, with an activity peak in late afternoon to early evening, and lowest activity level at 0000–0600 hours (Laughrin, 1977). A field study on Santa Cruz Island showed a distinct seasonal influence on activity patterns. In summer, activity was mainly crepuscular, while winter brought a marked increase in diurnal activity (Fausett, 1982).

REPRODUCTION

Courtship begins in January, and mating occurs in mid-March. Gestation is approximately 50–53 days. (Mainland gray foxes have a gestation of 53 days.) Pups are born from the end of April to early May. Mean litter size for 24 dens was 2.17 pups; the largest litter was five pups (Laughrin, 1977). Sex ratios on Santa Cruz Island are 1:1 (n = 275) (Laughrin, 1977). Pups remain with the family group until fall, when the families fragment and the young disperse. Yearlings sometimes disperse long distances (Laughrin, 1977).

Island gray foxes are unusually disease-free. Blood samples from 100 individuals indicated a complete lack (and lack of exposure to) rabies, distemper, and leptospirosis (Laughrin, 1977).

SOCIAL ORGANIZATION AND BEHAVIOR

Mated pairs remain together from January to May when the pups are born, and family units are together from May to September. Although pairs have been observed throughout the year, island gray foxes are usually solitary from September to December (Laughrin, 1977). On rare occasions groups larger than family units have been observed, but these aggregations occur only at unnatural food sources. Groups of 15–20 foxes congregated at a mess-hall-scrap dump site on San Nicolas Island. In this situation, foxes avoided each other but no agonistic interactions were observed (Laughrin, 1977).

Laughrin (1977) found that population densities varied considerably within and between islands, seeming to depend on habitat diversity and food availability. Santa Catalina, however, has an anomalously low population density, for no apparent reason. Population densities range from 0.3 foxes per km^2 on Santa Catalina to 7.9 foxes per km^2 on Santa Cruz. Home ranges are approximately 0.32 km^2 (Laughrin, 1977). On Santa Cruz Island home ranges of males and females are roughly equivalent during the summer. In winter, males' home ranges are roughly twice the size of females' (Fausett, 1982). This seasonal shift, also present in mainland gray foxes, reflects the females' restriction to den sites during breeding and pup-rearing seasons. Island gray foxes (on Santa Cruz) generally restricted their movements to their home ranges, but there was often such a high degree of range overlap that the concept of strict territoriality did not apply (Laughrin, 1977). Compared with mainland gray foxes, island gray foxes have higher population densities, "smaller home ranges, overlapping use of areas, [and] increased proportion of older animals in the population." These differences are related to their island habitat with its lower levels of high-risk mortality factors, and reduced competition (Laughrin, 1977, p. 43). The historical record indicates that considerable population fluctuations have occurred.

Laughrin (1977) observed frequent scent marking. Conspicuous objects along travel routes were marked, and scats were almost always placed "in open conspicuous sites or upon other objects or other scats" (Laughrin, 1977, p. 26). Laughrin observed mutual grooming between paired adults and an adult and juvenile. He also noted the use of the inverted-U tail position. Fighting, accompanied by growling, barking, and chasing, occurred often.

Corsac Fox (*Vulpes corsac*)
Credit: Tom McHugh/National Audubon Society
Collection/Photo Researchers, Inc.

CHAPTER 15
Genus *Vulpes*

Vulpes bengalensis: Bengal Fox

Bengal foxes are common throughout the Indian subcontinent. They are adaptable and small, traits that are at least partially responsible for their continued wide dispersal. Johnsingh's (1978) study of behavior and ecology is the most extensive research on free-ranging Bengal foxes available.

DISTRIBUTION AND HABITAT

Bengal foxes are found throughout the Indian subcontinent from its southernmost regions up into Nepal and the Indian state of Assam on the east and through Pakistan to the west. They exist in the foothills of the Himalayas up to 1,500 m. There are no records of populations in Afghanistan or Iran (Roberts, 1977; Mitchell, 1977; Johnsingh, 1978). These foxes inhabit both scrubby and open habitats, including alluvial plains and agricultural areas. They do not occupy heavily forested areas. They dig burrows, which may be used for several years (Roberts, 1977; Mitchell, 1977; Clutton-Brock *et al.*, 1976).

PHYSICAL CHARACTERISTICS

Bengal foxes are medium-sized and typically vulpine in appearance. Adult males weigh from 2.7 to 3.2 kg; females weigh under 1.8 kg (Roberts, 1977). [However, Acharyjo and Misra (1976) give the weight of a captive female as 2.4 kg.] Head-plus-body length is 45–60 cm. Tail length is 25–35 cm (Roberts, 1977). The pelage is short and smooth, and ranges in color from buff to silver-gray with an overall grizzled effect. The dorsal region is darker than the rest. The limbs are tawny and the underparts light, a pale sand or ginger shade. The tail is proportionately long, more than half the head-plus-body length, and is less bushy than that of *V. vulpes* or *V. cana*. It is black tipped, and may have a dark patch over the caudal gland. The muzzle is pointy, and there may be a dark smudged marking along the upper muzzle in front of the eyes. The backs of the ears are dark brown to gray anteriorly, and white inside (Roberts, 1977; Mitchell, 1977; Clutton-Brock et al., 1976; Mivart, 1890). See Roberts (1977, p. 101) for comparative drawings. The skull is typically vulpine with long, sharply pointed canines. The molars are well developed (Clutton-Brock et al., 1976).

TAXONOMY

According to the analysis of overall characteristics by Clutton-Brock et al. (1976), the Bengal fox is the "most typical" member of the *Vulpes* genus. These authors stated that it is reasonable to assume that "*Vulpes bengalensis* typifies the `basic fox'" (Clutton-Brock et al., 1976, p. 155). Stains (1975) noted that V bengalensis has been regarded as a subspecies of both V. vulpes and V. corsac. Currently, however, specific status seems warranted (Clutton-Brock et al., 1976; Nowak and Paradiso, 1983; van Gelder, 1978). *Vulpes bengalensis* is monotypic (Stains, 1975).

DIET

Bengal foxes are omnivorous. They rely primarily on insects, small mammals, ground-nesting birds, and vegetable material. Johnsingh's (1978) analysis of scats of free-ranging Bengal foxes in Tamil Nadu (southeastern India) showed a reliance on beetles, grasshoppers, crabs, lizards, field rats, and field mice. Some remains of scorpions, ants, termites, and spiders were also found in scats. Parts of rat snakes, hedgehogs, and various birds were found around dens. Mitchell (1977) observed free-ranging Bengal foxes in Nepal hunting along rice paddies and hedgerows, taking insects and small rodents. Other reports mention a reliance on fruit, such as wild melons, ripe fruit of banyan trees (*Ficus benghalensis*), berries, shoots and pods of *Cicer arietum*, and ber fruit (*Zizyphus*) (Roberts, 1977; Mitchell, 1977; Johnsingh, 1978; Mivart, 1890; Clutton-Brock et al., 1976). Frogs, bird eggs, beetle grubs, and moths are also mentioned as foods of free-ranging Bengal foxes (Johnsingh, 1978; Roberts, 1977; Mivart, 1890; Clutton-Brock et al., 1976). A captive thrived on an almost meatless diet of plantains, rice, and milk (Webb-Peploe, 1947, cited in Roberts, 1977).

ACTIVITY

Members of this species are generally nocturnal (Roberts, 1977; Mitchell, 1977), or crepuscular and nocturnal (Johnsingh, 1978). They may occasionally be active through the daylight hours as well (Mitchell, 1977).

REPRODUCTION

Mating occurs in December and January. The young are born in February and April, after a gestation period of 50–53 days. Litters usually number four (Roberts, 1977; Mitchell, 1977; Asdell, 1964, cited in Acharjo and Misra, 1976). Acharjo and Misra (1976) observed a copulatory tie that lasted several minutes in a pair of captives. On one occasion Johnsingh observed two different females sequentially suckling four pups. The parentage of the pups was unknown. There seems to be a persistent pair bond, and Johnsingh also observed a male playing with 2–4-month-old pups.

Humans and domestic dogs are a major mortality factor for free-ranging Bengal foxes in the Tamil Nadu area (Johnsingh, 1978). There is no information on longevity in this species.

SOCIAL ORGANIZATION AND BEHAVIOR

Little is known about the social organization of Bengal foxes. They seem to hunt alone. Johnsingh (1978) reported seeing a group of five almost full-grown foxes resting together under a rock.

Bengal foxes are not particularly wary of humans and can be found near human habitation. They are readily tamed (Müller-Using, 1975a; Mivart, 1890; Bueler, 1973). Nothing is known about territorial or scent-marking behaviors. They do not practice site-specific defecation. Vocalizations include whines, growls, and an alarm call characterized as a "chattering cry" (Johnsingh, 1978; Mivart, 1890). Yapping or baying in response to humans has also been observed (Mitchell, 1977).

Vulpes cana: Blanford's Fox

Common names for this species include hoary fox, Afghan fox, Baluchistan fox, and king fox (Roberts, 1977; Ognev, 1962; Novikov, 1962; Stains, 1975; Clutton-Brock *et al.*, 1976). Apart from descriptions of physical characteristics and Roberts's notes on diet, nothing is known of its natural history or behavior.

DISTRIBUTION AND HABITAT

Blanford's foxes are found in a small area of Asia. Their distribution is not well documented. They occur from northeastern Iran, where they are rare, to Afghanistan and northwestern Pakistan (Roberts, 1977; Novikov, 1962; Ognev, 1962; Clutton-Brock et al., 1976; Stains, 1975). Recently, Blanford's foxes have been found in Israel and the Sinai (Ilany, 1983; Mendelssohn et al., 1987), and two specimens from Oman have been described as well (Harrison and Bates, 1989). There is also a record of a possible sighting in the Rann of Kutch (northwestern India east of the Arabian Sea, south of Baluchistan) (Ranjitsinh, 1985).

These foxes are confined to mountain-steppe habitat in areas of barren rocky hills, interspersed with stony plains and small patches of cultivation. They prefer steep cliffs and rocky terrain (Mendelssohn et al., 1987), but they are not found in high mountain ranges (Roberts, 1977).

PHYSICAL CHARACTERISTICS

These are small foxes, weighing no more than 3 kg, with a head-plus-body length of 40–50 cm, a proportionately long tail of 30–41 cm, and ears of 6.5–7.4 cm (Roberts, 1977; Novikov, 1962). Foxes from Israel all weighed less than 1.5 kg (Mendelssohn et al., 1987). The pelage, for which this species has been and is still hunted, is dense, soft, and luxuriant. Although variation exists, the basic color ranges from straw-gray to quite dark, and there may be a blackish-brown zone running down the spine. Overall, members of this species are grayer than corsac foxes (Ognev, 1962; Roberts, 1977; Novikov, 1962; Clutton-Brock et al., 1976). The flanks are light straw-gray to buff, markedly lighter than the back. The legs are ocher to buff, though the hind legs may be darker. The toe pads are not furry, a characteristic that differentiates them from Rüppell's foxes, V. rüppelli (Mendelssohn et al., 1987). The throat, chest, and underparts are whitish-straw to white, and the fur on these parts is dense, long, and silky (Roberts, 1977; Ognev, 1962; Clutton-Brock et al., 1976). The muzzle, slender and delicate, is a grizzled gray, and a black band extends from the eyes to the upper part of the muzzle. Both the muzzle and the ears are pointy. Dorsally, the ears continue the color of the back; inside they are fringed with white (Roberts, 1977; Ognev, 1962; Novikov, 1962). The tail is luxuriantly furred and is almost equal to the body in length. Generally straw-gray in color, it is flecked by black-tipped hairs. There is a distinct brown–black caudal gland spot, and usually a dark tip to the tail, although some specimens may have white hairs at the extreme tail tip (Roberts, 1977; Novikov, 1962; Ognev, 1962; Clutton-Brock *et al.*, 1976).

The skull is quite small and delicate (Harrison and Bates, 1989). It is markedly smaller than that of *V. corsac*, and condylobasal length exceeds that of the fennec fox (*Fennecus zerda*), the smallest member of the family Canidae, by only a few millimeters (Clutton-Brock *et al.*, 1976; Ognev, 1962). The range of skull measurements for *V. cana* overlaps only slightly with that of the larger *V. rüppelli* (Mendelssohn *et al.*, 1987). The skull of *V. cana* is distinguished by its relatively delicate and slender rostrum (Harrison and Bates, 1989). The skull and dentition are typically vulpine. The nasal bones have a narrow wedge form, and the small teeth are sharply pointed (Clutton-Brock *et al.*, 1976; Novikov, 1962). See Dayan *et al.* (1989) for a discussion of character displacement in sympatric Blanford's foxes, Rüppell's foxes (*V. rüppelli*), and red foxes (*V. vulpes*).

TAXONOMY

The analysis of Clutton-Brock *et al.* (1976) showed little similarity between *V. cana* and all other members of the genus *Vulpes*. Blanford's fox most closely resembles Rüppell's fox (*V. rüppelli*), and in skull and dental characteristics it is most similar to fennec foxes (*Fennecus zerda*) (Clutton-Brock *et al.*, 1976, pp. 155–157).

V. cana is monotypic (Stains, 1975).

DIET

According to reports of local people in Baluchistan, Blanford's foxes rely on fruit for part of their diet, consuming melons, cultivated grapes, and fruits of the Russian olive (*Eleagnus hortensis*) (Roberts, 1977). Roberts (1977) remarks that these foxes probably eat locusts, lizards, and rodents as well. According to the scat analysis by

Mendelssohn *et al.* (1987), Blanford's foxes feed on arthropods, beetles, grasshoppers, scorpions, and small mammals. In Israel, scat analysis shows an insectivorous diet (Ilany, 1983).

ACTIVITY

In Israel, Blanford's foxes are nocturnal (Ilany, 1983). No other information on activity patterns is available.

REPRODUCTION AND SOCIAL ORGANIZATION AND BEHAVIOR

Ilany (1983, p. 150) commented that "observations on behaviour in the wild and in captivity show some similarity between this fox and the fennec, *Fennecus zerda*." He did not elaborate on this. No information on reproduction or social organization is available.

Vulpes chama: Cape Fox

Common names for this species include silver jackal, silver fox, and Cape fox (Bothma, 1966, 1971a; Roberts, 1951; Dorst and Dandelot, 1969; Ewer, 1973; van der Merwe, 1953a). These foxes seem to be common in southern Africa but, with the exception of studies by Bothma (1966, 1971a), no research on their life history or behavior has been undertaken. No records on the behavior of captives are available.

DISTRIBUTION AND HABITAT

The Cape fox is the only species in the genus *Vulpes* that exists in Africa south of the equator. The range originally encompassed the arid and semi-arid western areas of southern Africa (Meester and Setzer, 1971). Currently it includes northern Cape Province, southern and central Namibia, Botswana, southwestern Angola, Zimbabwe, and the Transvaal (Meester and Setzer, 1971; Clutton-Brock *et al.*, 1976; Shortridge, 1934; van der Merwe, 1953a; Bueler, 1973). According to Shortridge (1934) the range does not extend north of the Zambezi River.

These are animals of arid environments. They are found mainly on open plains and karoo veldt, as well as on the Kalahari savanna (Meester and Setzer, 1971; Dorst and Dandelot, 1969; Shortridge, 1934; van der Merwe, 1953a). Shortridge (1934) commented that Cape foxes prefer habitat at the bases of kopjes and along stony ridges. They are never found in forested areas.

PHYSICAL CHARACTERISTICS

These are medium-sized foxes, similar in size to Bengal foxes, *V. bengalensis*. Weights range from 3 to 4.5 kg. Head-plus-body length is about 56 cm, height at shoulder is 30 cm, and tail length roughly 33 cm (Dorst and Dandelot, 1969; van der Merwe, 1953a; Clutton-Brock *et al.*, 1976; Nowak and Paradiso, 1983). Pelage color is silvery-gray due to mixing of black and white guard hairs. It is generally soft and short, and paler in some individuals than in others (Roberts, 1951; Clutton-Brock *et al.*, 1976). The neck and sides are lighter in color, and the underparts are pale tawny to pale buff. The head is dull red and the lower jaw dark brown. The ears are relatively large, their backs tawny with white hairs in front of them. There is a white marking on the throat. The legs are tawnier than the rest of the pelage (Roberts, 1951; Dorst and Dandelot, 1969; van der Merwe, 1953a). The tail is thick and bushy, and may be silvery, pale fawn, buff with brown- or black-tipped hair, or dull yellow. The tail tip is always entirely black, and there is a dark patch over the caudal gland (Dorst and Dandelot, 1969; Roberts, 1951; van der Merwe, 1953a; Clutton-Brock *et al.*, 1976).

The muzzle is short and pointy (Dorst and Dandelot, 1969). The skull is very similar to that of *V. bengalensis*, although the cranium of *V. chama* is slightly wider and the maxillary region is slightly shorter (Clutton-Brock *et al.*, 1976).

TAXONOMY

V. chama is one of the three *Vulpes* species in Africa (the other two are *V. pallida*, the pale fox, and *V. rüppelli*, Rüppell's fox). Clutton-Brock *et al.* (1976) found a very close similarity to *V. bengalensis* and a lesser, though still high, degree of similarity to *V. pallida*. No subspecies have been described.

DIET

The most detailed information available on any aspect of the natural history of Cape foxes is from stomach-content analyses of wild-caught individuals (Bothma, 1966, 1971a). These reports, combined with others of a less detailed nature, show Cape foxes to be completely omnivorous. Bothma's studies show a diet composed of the following items (listed in decreasing order of importance): rodents, carrion (not of domestic hoofstock), insects, and vegetable material. Other dietary items are birds, reptiles, lagomorphs, and cultivated crops. Other food lists include gerbils; field mice and other small rodents; hares; birds; bird nestlings and eggs; diverse vegetable material, including wild fruit, berries, seeds, roots, and tubers; lizards; insects, such as white ants, beetles and their larvae, and locusts; and carrion (Meester and Setzer, 1971; Dorst and Dandelot, 1969; Shortridge, 1934; Roberts, 1951; Nel, 1984; van der Merwe, 1953a). Meester and Setzer (1971) remarked that insects seem to constitute a major part of Cape foxes' diets. Cape foxes scavenge around areas of human use (and refuse) (Roberts, 1951; van der Merwe, 1953a). They also cache food (Nel, 1984).

Bothma's studies overrepresented specimens from regions of heavy sheep farming, but he found no evidence of domestic stock in any of the 69 stomachs examined (Bothma, 1971a). This finding should help dispel the notion that Cape foxes kill domestic stock. They may kill domestic fowl, though even this habit is debatable (van der Merwe, 1953a; Shortridge, 1934).

ACTIVITY

Cape foxes are primarily nocturnal with occasional crepuscular activity (Meester and Setzer, 1971; Dorst and Dandelot, 1969; Nel, 1984; Shortridge, 1934; van der Merwe, 1953a). Hunting is strictly nocturnal. During daylight hours these foxes remain hidden under rocks or in their burrows (Dorst and Dandelot, 1969), although pups may play outside during daylight hours (Nel, 1984).

REPRODUCTION

The breeding season begins in August or September, and the gestation period is 7–8 weeks (Roberts, 1951). There are three to five young per litter (Meester and Setzer, 1971; Dorst and Dandelot, 1969; van der Merwe, 1953a). Cape foxes use burrows dug by other animals (van der Merwe, 1953a). The duration of the pair bond and its stability from year to year are unknown. Some male parental care has been observed, but how long the male remains with the family group is unknown. Males provision postparturient females for at least a week or two (Nel, 1984). Multiple litters at a den have been observed (Nel, 1984). Some Cape fox subadults steal food brought to the pups at the den by the parents (Nel, 1984).

SOCIAL ORGANIZATION AND BEHAVIOR

Almost nothing is known about the social organization of Cape foxes. They seem to be solitary or sometimes to associate in pairs (Meester and Setzer, 1971; Dorst and Dandelot, 1969; Shortridge, 1934). They appear to forage and feed singly, even when the male is attached to the family group (Nel, 1984). The extent to which pairs associate outside the breeding season is unknown. In the Orange Free State, home ranges of Cape foxes may overlap (Nel, 1984).

Vocalizations include growls and barks. A long-range vocalization of yelps or yapping barks has been described, but Cape foxes apparently do not howl (Shortridge, 1934; van der Merwe, 1953a; Dorst and Dandelot, 1969).

Vulpes corsac: Corsac Fox

Intensively hunted for fur, killed inadvertently in marmot traps, and threatened by settlement of the steppes on which they live, corsac foxes are now much reduced in number (Stroganov, 1962; Ognev, 1962; Nowak and Paradiso, 1983). Despite their economic importance and wide distribution, little is known about the life history and habits of this species. Sidorov and Botvinkin (1987) published a detailed study of corsac fox distribution and ecology in southern Siberia (Russian, with a brief English abstract).

DISTRIBUTION AND HABITAT

Corsac foxes are widely distributed in Asia. They occur from the Azov Sea (the northern arm of the Black Sea) west through northeastern

China and Mongolia to the Transbaikalian Steppes into northern Manchuria. Some populations occur in northeastern Iran and northern Afghanistan (Ognev, 1962; Stroganov, 1962; Novikov, 1962; Clutton-Brock et al., 1976; Stains, 1975).

These are animals of steppes and subdesert zones. They prefer rolling semi-desert habitats and avoid heavy forests, brushy areas, and regions of heavy human habitation or agricultural use. In years when corsac fox populations are high or when food is scarce in their usual range, some foxes move into forested steppe habitats. Their range extends to an elevation of 300 m (Ognev, 1962; Novikov, 1962; Stroganov, 1962). Corsac foxes inhabit burrows dug by marmots, badgers, and other foxes, although sometimes they may dig burrows of their own (Novikov, 1962).

PHYSICAL CHARACTERISTICS

Corsac foxes are about two thirds the size of common red foxes (*V. vulpes*) and roughly the same size as arctic foxes (*Alopex lagopus*). They look typically vulpine except for proportionately longer legs, which may be an adaptation to the open steppes, and large broad ears. Head-plus-body length is 50–60 cm, with a shoulder height of about 30 cm, tail length of 25–35 cm, and an ear length of 8 cm. Some individuals may weigh as little as 2.3 kg (Novikov, 1962; Ognev, 1962; Stroganov, 1962; Bueler, 1973).

Pelage color varies with the seasons, but the general effect is of fawn to straw-gray, with the winter coat a paler straw-yellow color. The area along the spine is slightly darker. In winter the fur is dense and soft, while in summer it is shorter and coarser and the tail is a lot less bushy. The flanks are light in color, the limbs are yellow-red laterally, yellow-white medially (Mivart, 1890). Chin, throat, chest, and underparts are very pale (yellowish to pure white). The muzzle is slender, pointy, and reddish-brown to straw-gray. The broad ears, more widely set than in *V. vulpes*, are yellow-gray to rufous-gray posteriorly, whitish-yellow to white anteriorly. The tail is straw-yellow with a paler underside. Its tip is brown or black, and there is a black caudal gland spot (Ognev, 1962; Novikov, 1962; Stroganov, 1962; Clutton-Brock et al., 1976; Mivart, 1890).

The skull is smaller than in *V. vulpes*, with a more gradually tapering rostrum and a weakly developed low sagittal crest. The auditory bullae are small and flat, and the teeth are small and weakly developed, the upper first molar in particular being much reduced in size (Novikov, 1962; Ognev, 1962; Clutton-Brock *et al.*, 1976; Stroganov, 1962).

TAXONOMY

The analysis by Clutton-Brock *et al.* (1976) showed a close resemblance to *V. bengalensis*, the Bengal fox. Ognev (1962) and Stains (1975) stated that there are three subspecies, while Stroganov (1962) listed four, two of which occur in Siberia.

DIET

Corsac foxes exhibit the dietary patterns typical of most vulpine species. They eat small mammals, such as voles, mice, pika, young marmots, hedgehogs, jerboas, ground squirrels, hares, and polecats. Other prey items include birds and their eggs, small reptiles such as lizards, frogs, insects, and carrion. They are not known to attack domestic fowl (Novikov, 1962; Ognev, 1962; Stroganov, 1962; Clutton-Brock *et al.*, 1976; Bueler, 1973). A stomach-content analysis of 18 corsac foxes in Transbaikalia (Brom *et al.*, 1948, cited in Stroganov, 1962) showed the following: rodents, 85%; insects, 22%; undetermined mammals, 16.5%; carnivores (polecats and other corsac foxes), 5.5%; birds, 5.5%. Seasonal changes in food consumption patterns are small (Novikov, 1962). Corsac foxes can exist for long periods without access to water (Novikov, 1962; Ognev, 1962).

ACTIVITY

Free-ranging corsac foxes seem to be nocturnal (Stroganov, 1962; Novikov, 1962). Captives at the Tierpark Berlin (in what was once East Berlin) were also active during the day (Dathe, 1966).

REPRODUCTION

Estrus occurs in January or February (Novikov, 1962; Ognev, 1962) and the young are born after a gestation period of 50–60 days (Novikov, 1962; Stroganov, 1962). Captives at the East Berlin Zoo had gestation periods of 49–51 days and reproduced readily (Dathe, 1966). Litter size ranges from two to 11 pups, with a usual range of 2–6 (Dathe, 1966; Ognev, 1962). The importance of the male with respect to pup rearing is unclear. In captivity, males may safely be left with their parturient mates, and thereafter they may play some part in rearing and defending the young (Dathe, 1966).

SOCIAL ORGANIZATION AND BEHAVIOR

Corsac foxes are more gregarious than other fox species (Ognev, 1962), although continued long-term hunting pressure from man may have exerted a strong selective pressure against their highly visible social organization. Sites with large numbers of simultaneously inhabited burrows known as "Corsac cities" used to be reported but now no longer exist (Novikov, 1962; Ognev, 1962). Small packs may hunt together (Dinnik, 1914, cited in Ognev, 1962; Stroganov, 1962), and several foxes may live in one burrow (Ognev, 1962). Whether these packs are parent-offspring groups or more complex social arrangements is unknown.

In captivity, corsac foxes are readily tamed and are quite sociable and playful (Dathe and Pedersen, 1975; Ognev, 1962). Vocalizations include whimpers, growls, and barks.

Vulpes ferrilata: Tibetan Sand Fox

Little is known about Tibetan sand foxes, a small member of the *Vulpes* group from the Himalayan region.

DISTRIBUTION AND HABITAT

Tibetan sand foxes occur in Tibet and in the Mustang District of northern Nepal (Mitchell, 1977; Clutton-Brock *et al.*, 1976; Stains, 1975). Bueler (1973) stated that they may occur in the Sutlej Valley of extreme northwestern India. These foxes live on high plateaus in alpine desert habitat at elevations of 3,000 m and above (Mitchell, 1977; Clutton-Brock *et al.*, 1976). In the Mustang District of northern Nepal, they live on barren slopes and streambeds. In this same region, where they are common, their tracks are abundant in new snow along stream banks and around wheat fields (Mitchell, 1977). Dens are in piles of boulders and in burrows under large rocks.

PHYSICAL CHARACTERISTICS

Tibetan sand foxes are medium sized in relation to the other nine species in the genus *Vulpes*. Head-plus-body length ranges from 57.5 to 70 cm with a tail length of 40–47.5 cm (Mitchell, 1977). Mivart (1890) gave a tail length of only 29.5 cm for the specimen he examined, and Stains (1975) remarked that their tails are proportionately shorter than those of other vulpines. These foxes are typically foxlike in appearance, the only remarkable feature being a particularly elongated rostrum (i.e. long snout) (Mitchell, 1977; Clutton-Brock *et al.*, 1976). Males weigh up to 7 kg (Mitchell, 1977). The ears are not as large relative to body size as in some of the other fox species, perhaps as an adaptation to the extreme climate. The pelage is thick, with long fur on the legs and feet (Mitchell, 1977; Mivart, 1890; Stains, 1975). The pelage on the back and sides of the body is pale to bright rusty-yellow, sandy, or pale gray agouti (Clutton-Brock *et al.*, 1976; Mitchell, 1977; Mivart, 1890). Mivart (1890) mentioned a distinct white area on the chest surrounded by gray fur. The insides of the ears and the tip of the very well-furred tail are white as well. There may be a dark spot or streak on the tail over the caudal gland area (Clutton-Brock *et al.*, 1976). The sides of the neck and body, as well as most of the tail, are a mixture of black and white or gray hairs, giving an overall gray appearance (Mivart, 1890; Mitchell, 1977). The limbs are yellowish-rufous to tawny on their fronts (Clutton-Brock *et al.*, 1976).

The skull of members of this species is elongate with a particularly narrow maxillary region. The canines are remarkably elongated and pointy, and the cheek teeth are widely spaced and well developed (Clutton-Brock *et al.*, 1976). The auditory bullae are quite well inflated (Mitchell, 1977).

TAXONOMY

Phenotypic analysis shows the closest similarity with *V. corsac*. The Tibetan sand fox may have evolved from this similar congeneric in response to its specialized environment (Clutton-Brock *et al.*, 1976). *V. ferrilata* is monotypic (Stains, 1975).

DIET, ACTIVITY, REPRODUCTION, AND SOCIAL ORGANIZATION AND BEHAVIOR

Almost nothing is known of the life history and habits of these foxes. They do not seem to be solitary. Mitchell (1977) observed pairs hunting together. He stated that they prey on rodents, lagomorphs, and ground-nesting birds. Bueler (1973) stated that they eat rockhares (black nosed pikas). Free-ranging individuals breed in late February, and two to five pups are born in April or May (Mitchell, 1977).

Vulpes macrotis: Kit Fox

These foxes are the smallest of the three species of *Vulpes* occurring in North America and are among the smallest of the vulpines worldwide. In many respects kit foxes are the North American counterparts of the fennec foxes (*Fennecus zerda*) of North Africa. Both species exhibit shared adaptations to the extreme arid conditions of the deserts they inhabit. As is also true for swift foxes (*V. velox*), kit fox populations have been drastically reduced in some areas due to human activities. Poisoning campaigns, shooting, trapping, and habitat destruction have all played their part. Like swift foxes, kit foxes have little wariness of traps and poison baits and are killed frequently, though usually inadvertently, in coyote-control efforts. (McGrew, 1977, 1979; Egoscue, 1956, 1962; Morrell, 1972). The subspecies *V. m. mutica* (San Joaquin kit fox) was classified as a federally protected endangered species in 1966, and as rare by the state of California (O'Farrell, 1984; Samuel and Nelson, 1982). A number of authorities consider *V. macrotis* a subspecies of *V. velox*, the kit fox (see *taxonomy* section below).

DISTRIBUTION AND HABITAT

Kit foxes occur in arid and semi-arid regions of the western United States and northwestern and north-central Mexico, including the Baja Peninsula. In the United States the range includes the extreme southwestern corner of Oregon and extends southward to portions of Idaho, Nevada, Utah, California, Arizona, and New Mexico (McGrew, 1979; Boynton, 1970; Egoscue, 1956; Samuel and Nelson, 1982; O'Farrell, 1987). See O'Farrell (1987) for a detailed distribution map.

Kit foxes occur in dry habitats, such as scrub desert. In Utah the majority of kit fox sightings are in areas of desert shrub at elevations below 1,600 m (McGrew, 1977). Dens or burrows are used year-round. These are often found in groups on preferred sites, with 8 or 10 dens within 0.8–1.2 ha. Most of these are not used simultaneously, and kit foxes may move from one den to another. Whelping dens are separated by distances of at least 3.2 km (Egoscue, 1962). Dens, particularly those used by family groups, have multiple entrances. Less complex small burrows are used as refuges when away from the home den (Egoscue, 1962; McGrew, 1979; Boynton, 1970). Kit foxes are not particularly wary of humans and may den within 50 m of a building (Egoscue, 1956, 1962). Although kit foxes are desert canids, they can swim; Reeder (1949) observed a kit fox attempting to ford a broad, swift-flowing canal.

PHYSICAL CHARACTERISTICS

Kit foxes are typically vulpine in appearance, and are slightly smaller than, but very similar to, swift foxes (*V. velox*). Kit foxes are the smallest members of the North American *Vulpes* species, averaging 25% smaller than red foxes (*V. vulpes*) in most linear measurements. They are thoroughly adapted to their desert existence, and share many of the characteristics of other arid-habitat vulpines (i.e., *V. rüppelli*, *V. cana*, *V. corsac*, *V. velox*, and *Fennecus zerda*), such as small size, light-colored pelage, large ears, and nocturnal habits (McGrew, 1979; Golightly and Omhart, 1984; Thornton and Creel, 1975; Turkowski, 1973). Pelage coloration is variable, but the essential tones are dusty grizzled-gray, yellow-gray, or buff-gray, with the shoulders, flanks, and chest ranging from buff to orange. Guard hairs are black tipped or banded with black and white. The underfur is heavy and somewhat coarse. The underparts are light buff to white. The thick fur acts as insulation against both heat and cold. The legs are slender and continue the color of the body. The soles of the feet are protected by stiff tufts of hair, a trait that may improve traction on loose sandy surfaces, as well as protecting against extremes of temperature (Boynton, 1970; Turkowski, 1973; McGrew, 1979; Grinnell *et al.*, 1937, cited in McGrew, 1979). The ears are huge, a trait common among desert-dwelling mammals. They are proportionately larger than in swift foxes (Zeveloff, 1988). Lined by a network of fine hairs, they are buff at their base. The muzzle and vibrissae are generally black to brown, although some individuals show no black or brown here. The tail is about 40% of the total body length, and thus is proportionately longer than the tails of swift foxes. It is bushy and gray, and there is a pronounced black spot over the caudal gland (McGrew, 1979; Turkowski, 1973). There is slight sexual dimorphism, with males averaging 2.2 kg, females 1.9 kg. Head-plus-body length ranges from 35 to 50 cm. Tail length is 22.5–32 cm. On average, males are considerably heavier than females, but in terms of external measurements, females are only very slightly smaller (McGrew, 1979). The young of the year reach adult size by mid-August (Morrell, 1972; McGrew, 1979; Egoscue, 1962).

The skull is delicate with a long, narrow rostrum and large auditory bullae. It is slightly longer overall, and the auditory bullae are larger, than in *V. velox*. On the basis of these and other characteristics, the skulls of *V. macrotis* and *V. velox* can be readily differentiated (Dragoo *et al.*, 1987; Creel and Thornton, 1971). There is slight, but statistically insignificant, sexual dimorphism in cranial and skeletal characteristics (McGrew, 1979; Thornton and Creel, 1975; Dragoo *et al.*, 1986, 1987). The dental formula is incisors 3/3, canines 1/1, premolars 4/4, molars 2/3 = 42.

TAXONOMY

Some authorities consider *V. macrotis* and *V. velox* as a single species (Dragoo *et al.*, 1990; Hall, 1981; Clutton-Brock *et al.*, 1976; Ginsberg and Macdonald, 1990; Ewer, 1973; van Gelder, 1978). Interbreeding between the two species does occur where their ranges overlap (eastern New Mexico and western Texas), but this hybridization is quite restricted in scope (Rohwer and Kilgore, 1973; Thornton and Creel, 1975; Thornton *et al.*, 1971). A number of external characteristics, as well as evidence from electrophoretograms of hemoglobin and serum proteins, argue for distinct specific status (McGrew, 1979). Thornton and Creel (1975), who completed a detailed study of the taxonomic issues involved, also stated that present evidence favors distinct specific status. The biochemical analyses of Dragoo *et al.* (1990), however, suggested a level of merely subspecific differentiation, while their morphometric data were more ambiguous. The karyotypes of kit and swift foxes are the same (2n=50) (Dragoo *et al.*, 1990).

Eight subspecies have been described (McGrew, 1977, 1979; O'Farrell, 1987; Snow, 1973a,b), but some of these descriptions are based on small numbers of specimens, and one subspecies is extinct, so this scheme is in need of revision. Waithman and Roest (1977) stated that there are only five valid subspecies. O'Farrell (1987) suggested that there may be as few as four. The San Joaquin kit fox (*V. m. mutica*) is listed as endangered by the U.S. Dept. of the Interior and as threatened by the state of California (Ginsberg and Macdonald, 1990).

DIET

Kit foxes are an essentially carnivorous species, although they consume insects and some vegetable material as well. Kit foxes rely primarily on small mammals, principally lagomorphs and rodents. Black-tailed jackrabbits represent the upper size limit of prey (Morrell, 1972; Egoscue, 1962; Turkowski, 1973; Fisher, 1981; McGrew, 1979). In western Utah kit foxes rely primarily on lagomorphs and secondarily on nocturnal rodents and birds. Items consumed in this region include black-tailed jackrabbits, kangaroo rats, deer mice and other mice, burrowing voles, cottontail rabbits, birds, and lizards. Remains at dens, and scats show that kit foxes eat horned larks and meadowlarks (Egoscue, 1956, 1962). There seems to be a correspondence between fluctuations in black-tailed jackrabbit and kit fox populations in Utah (Egoscue, 1975). San Joaquin kit foxes (in California) subsist primarily on kangaroo rats, secondarily on rabbits. They also eat gophers, pocket mice, ground squirrels, birds, lizards, scorpions, insects, vegetable material, and assorted roadkill (Morrell 1972). Insects consumed include grasshoppers and crickets. Kit foxes also eat cactus fruits (Egoscue, 1956; Morrell, 1972; McGrew, 1979; Turkowski, 1973; Boynton, 1970). Kit foxes may cache food (Morrell, 1972; Turkowski, 1973). They do not need access to water, and they may den some distance from any water source (Morrell, 1972; Golightly and Omhart, 1984; Boynton, 1970; McGrew, 1979).

ACTIVITY

As is true for most other desert-dwelling canids, kit foxes are nocturnal, a strategy allowing avoidance of desert daytime heat. Usually they remain inactive in their dens during the day, but they sometimes emerge to bask and play in the immediate vicinity of the den, particularly when the pups are young. Occasionally they hunt during daylight hours, particularly on overcast days, though for the most part hunting is strictly nocturnal (Morrell, 1972; Egoscue, 1956; Turkowski, 1973; Boynton, 1970; McGrew, 1979).

REPRODUCTION

Kit fox females are annually monestrous. The majority do not breed in their first year, although some do (O'Farrell 1987). The usual mating pattern of kit foxes is monogamy (but see below), and the mated pair remains together throughout the year. There is no evidence from field studies that kit foxes mate for life (Egoscue, 1956, 1962; Morrell, 1972; Turkowski, 1973; McGrew, 1979). Mating occurs anytime from December to February (McGrew, 1979; Egoscue, 1956, 1962). Gestation is 49–55 days (O'Farrell, 1987). Litters are born in February, March, or early April. Litter size ranges from two to six pups; the usual number is four or five (Egoscue, 1956, 1962; Morrell, 1972; Boynton, 1970; McGrew, 1979). Egoscue (1975) found that in Utah the average litter size covaries with the prey base, and that when kit fox populations are at the carrying capacity of an area, a preponderance of males is born.

Although the usual mating pattern seems to be monogamy, Egoscue (1962) observed three instances of polygyny. In each of these cases one of the females appeared to be younger than the other, and their litters were also of different ages. Morrell (1972) also stated that one male kit fox apparently mated with three females in California. Pairs or groups of three adults captured together by Egoscue (1975) were almost invariably composed of one male plus females. Unpaired adults never den together. Tracks observed in the snow in Utah in December and January indicated that kit foxes visit unoccupied dens and den sites— perhaps scouting for available dens or searching for mates (Egoscue, 1956). Egoscue (1956) stated that the monogamous pair bond lasts indefinitely, or until one of the pair dies. Contrastingly, of seven breeding pairs observed in California, only one consisted of the same two individuals during successive breeding seasons (Morrell, 1972). Turkowski (1973) stated that males usually pair with different mates each year. Obviously the question of long-term pair bonding in this species needs further research.

Males share in parental care. They provision their mates before the pups are weaned, and after weaning both parents hunt together to provision their young. They carry food back to the den for their pups

(Morrell, 1972; Turkowski, 1973; McGrew 1979; Boynton, 1970). The male may not stay in the same den with the female and pups, although he usually dens nearby. The pups begin to forage with their parents at the age of 3–4 months, and by October they are almost fully grown. The family group breaks up no later than October. The pups sometimes remain together for a while after the family breaks up. One or more offspring may not disperse, instead remaining behind at the den with the parents (Egoscue, 1956; Morrell, 1972; Turkowski, 1973; McGrew, 1979; Boynton, 1970).

Captives have lived over 20 years (Nowak and Paradiso, 1983). Egoscue (1975; also cited in McGrew, 1979) recorded that a 7-year-old free-ranging individual was "very feeble" with worn and broken teeth. Human-caused deaths seem to be the chief cause of mortality in many regions. Golden eagles, rough-legged and red-tailed hawks, and coyotes all prey on kit foxes (Egoscue, 1956, 1962, 1975; McGrew, 1979; O'Farrell, 1987).

SOCIAL ORGANIZATION AND BEHAVIOR

The basic social unit is the mated pair, who either remain together in the fall after their offspring have dispersed or separate until the next mating season. It seems that individuals can either mate with each other again the following season, or find new mates. Paired foxes hunt in the same area but not in any organized fashion (Morrell, 1972; Turkowski, 1973; Egoscue, 1956, 1962, 1975; McGrew, 1979). Pups disperse in the fall, and Egoscue (1962) recorded that, of 36 tagged pups that were followed, none ever became a resident of the original study area. Although dispersal distances are unknown, Egoscue (1956) stated that a tagged pup was recaptured 32 km from its natal den area.

Kit foxes do not seem to be particularly territorial. In California home ranges largely overlap, and no specific hunting territories are defended (Morrell, 1972). Foxes from different family groups hunt in the same areas but not simultaneously. In Utah, natal dens are always at least 3.2 km apart (Egoscue, 1962). Most dens there are less than 1.6 km from vegetated dunes, which support high rodent populations (Egoscue, 1956). Seventy-eight percent of transients counted in Utah were males (Egoscue, 1975).

Various estimates of population density have been published. In western Utah, population densities ranged from 1 adult per 470.9 ha to

1 adult per 1,035.9 ha (Egoscue, 1975). A maximum population density in optimum habitat may be 1.8–2.0 adults per 258.9 ha (Egoscue, 1975). In California Morrell (1972) found that optimal habitat supported two adults per 259 ha, a figure remarkably similar to Egoscue's estimate. In a Sonoran Desert habitat, home range was closer to 1,120 ha (Zoellick, 1985, cited in O'Farrell, 1987). Egoscue suggested that above a certain point the maximum carrying capacity of kit fox habitat is not tied to increasing prey population. Continued increases in rabbit populations beyond what seem to be required to support a maximum fox population do not result in continued fox population increases. Thus, Egoscue suggested, the carrying capacity of kit fox habitat is related to social requirements.

Kit foxes do not seem to use scent marking as a territorial advertisement. In Utah, scats were deposited on trails and near objects "such as bits of bone or other animal remains," but no regularly visited scent stations were discovered (Egoscue, 1962, p. 488). Scats are deposited near dens and within them; tunnels in a den may be used for latrines and refuse.

Short-range vocalizations include whimpers and purrs. Growls and barks occur in contexts of threat and warning. A distinctive "croaking noise" is emitted by trapped or cornered individuals. The long-range "lonesome call" seems to function in communication between separated adults or pups (Egoscue, 1962; Turkowski, 1973; Morrell, 1972).

Vulpes pallida: Pale Fox

As is true for more than half of the ten species of *Vulpes*, almost nothing is known about the natural history of *V. pallida*. Commonly known as pale, pallid, or sand foxes, these animals are not confined to sandy habitats and in some areas are not at all pale in color. Thus these names are not particularly apt.

DISTRIBUTION AND HABITAT

V. pallida occurs in a broad swath across the Saharan and sub-Saharan regions of the African continent from Senegal and Mauritania on the west coast, through Mali, Niger, Nigeria, northern Cameroon, Chad, and the northern provinces of the Sudan (Rosevear, 1974; Meester and

Setzer, 1971; Clutton-Brock *et al.*, 1976). These are animals of arid habitats that prefer savanna and open sandy areas (Dorst and Dandelot, 1969; Stains, 1975; Rosevear, 1974). Unlike Rüppell's foxes (*V. rüppelli*), who live in both sandy and rocky deserts, pale foxes are confined to nonrocky habitat. Members of this species may occur in slightly moister wooded habitats as well (Rosevear, 1974).

PHYSICAL CHARACTERISTICS

Pale foxes are medium sized, about the same size as Rüppell's foxes (*V. rüppelli*). Head-plus-body length is about 46 cm, tail length is 25–35 cm, shoulder height is 25 cm, and weight is 1.5–3 kg (Dorst and Dandelot, 1969; Rosevear, 1974; Bekoff, 1975). Pelage colors vary from buff to pale red, and are finely speckled. There is a darker band running along the spine. The flanks are lighter in color, and the underparts are buffy white, white, or occasionally reddish. The throat is white or reddish (Rosevear, 1974; Dorst and Dandelot, 1969). The head and back of the neck are generally the same color as the back. The cheeks are pale (pure white on some individuals), and the eyes are bordered by a darker ring. The muzzle is sharp and pointed. The ears are proportionately smaller than those of other desert fox species (*V. rüppelli, Fennecus zerda*). They are dorsally darker than the back and head, and white inside. The limbs are rusty to dark red-brown outside, lighter to white on their inner surfaces. The tail, long and well furred, is essentially the same color as the back, and it is flecked with black-tipped hairs. There is a dark marking over the caudal gland, and a very pronounced black tip (Rosevear, 1974; Dorst and Dandelot, 1969; Clutton-Brock *et al.*, 1976).

The skull is similar to that of *V. rüppelli*, although the auditory bullae of *V. pallida* are slightly larger and the nasals are longer. The carnassials are, however, shorter than those of *V. rüppelli* (Rosevear, 1974). In contrast with the rather weak carnassials, the molars are well developed (Clutton-Brock *et al.*, 1976).

TAXONOMY

Pale foxes have been divided into five subspecies. Rosevear (1974) pointed out that although individuals from different localities are dissimilar, there are no clear lines of distinction between variant forms.

Rosevear stated that these subspecific classifications are pointless. Stains (1975) stated that *V. pallida* is monotypic. Physically, pale foxes are most similar to Cape foxes (*V. chama*) (Clutton-Brock *et al.*, 1976).

DIET

Pale foxes feed on rodents, ground-nesting birds and their eggs, reptiles, and vegetable material, such as wild fruit (Meester and Setzer, 1971; Dorst and Dandelot, 1969). Pale foxes may catch and eat domestic fowl as well (Müller-Using, 1975b). Some zoo-kept captives ate mealworms, mice, and biscuits (Bueler, 1973).

ACTIVITY

Pale foxes are primarily nocturnal (Meester and Setzer, 1971; Nowak and Paradiso, 1983).

REPRODUCTION

No information on reproduction in free-ranging pale foxes is available. A captive female held at Nuremberg Zoo gave birth to a litter of four (Bueler, 1973). Another captive had a lifespan of three years (Rosevear, 1974).

SOCIAL ORGANIZATION AND BEHAVIOR

Pale foxes are gregarious. They live in communal burrows, probably in family groups (Meester and Setzer, 1971; Dorst and Dandelot, 1969). Rosevear (1974) described the species as fairly social and stated that numerous colonies of up to 30 burrows exist in the Sahel. Pale fox burrows are extensive, their galleries reaching 9–15 m (Dorst and Dandelot, 1969). Bueler (1973) reported that three adults (one female and two males) housed together in captivity got along well, even after the female gave birth to a litter of four.

Vulpes rüppelli: Rüppell's Fox

Commonly known as the sand fox or Rüppell's fox, *V. rüppelli* is one of the three species of the genus *Vulpes* found in Africa. Since the name "sand fox" is sometimes used to refer to the group of foxlike species adapted to arid habitats (*V. chama, V. pallida, V. rüppelli, Fennecus zerda*), the common name of Rüppell's fox is used here. Lindsay and Macdonald (1986) have completed the only field study of this species.

DISTRIBUTION AND HABITAT

Rüppell's foxes are found in arid regions of northern Africa, the Arabian Peninsula, and into Asia. Specifically, they are found from Algeria, central Niger, Libya, northern Chad, and Egypt, south to the Sudan and the arid lowlands of Ethiopia and northern Somalia. They are found throughout the Arabian Peninsula and north into Israel, including the Negev. Their range extends eastward into Iraq, which constitutes the northern limit of their distribution, and eastward into Afghanistan and Pakistan (Roberts, 1977; Harrison, 1968; Yalden *et al.*, 1980; Meester and Setzer, 1971; Rosevear, 1974). These foxes are highly adapted to life in arid environments. They live in rolling sand dune and rocky desert habitats (Roberts, 1977; Harrison, 1968; Dorst and Dandelot, 1969; Rosevear, 1974). Lewis *et al.* (1964, cited in Harrison, 1968) noted that in northern Saudi Arabia they are generally found along wadis.

PHYSICAL CHARACTERISTICS

Rüppell's foxes are distinguished by their large ears and slight build; otherwise they have the "foxy" look typical of all members of the genus *Vulpes*. Head-plus-body length ranges from 48 to 52 cm. Tail length is 25–35 cm. Height at shoulder is about 25 cm (Roberts, 1977; Dorst and Dandelot, 1969). Weight is 1.5–4 kg (Lindsay and Macdonald, 1986; Dorst and Dandelot, 1969; Müller-Using, 1975b). The slight build of Rüppell's foxes is indicated by the French common name *renard famélique*, which means "famished fox." The pelage is dense and soft. It is gray to buff with a silver cast due to numerous white hairs. There is a good deal of color variation among members of this species (Harrison, 1968; Rosevear, 1974). There is a deep cinnamon-rufous band along the spine, and the flanks are lighter. The underparts are usually creamy to white, although they may also be an intense red. The legs are rufous to rufous-white on the outside, white along their inner parts. The neck, front of the face, and muzzle are pale reddish to buff, and there is a dark marking extending from the eyes to just forward of the middle of the upper lip (Dorst and Dandelot, 1969; Roberts, 1977; Rosevear, 1974). The large ears are white anteriorly; their posterior surface is a deep cinnamon-rufous to pale red. The long tail is very well-furred, buff with long black-tipped hairs and a white tip. There is a dark marking at the caudal gland. The soles of the feet are covered in long, soft hairs that conceal the pads completely, an adaptation to extremes of temperature in the desert also shared by the fennec fox (*Fennecus zerda*). As in other desert foxes, the facial vibrissae are particularly long (Clutton-Brock *et al.*, 1976). See Dayan *et al.* (1989) for a discussion of ecological character displacement in sympatric Rüppell's foxes, red foxes (*V. vulpes*), and Blanford's foxes (*V. cana*).

TAXONOMY

Stains (1975) remarked that *V. rüppelli* is sometimes considered to be a subspecies of *V. corsac*, but this no longer seems to be a widely held view (Clutton-Brock *et al.*, 1976; Meester and Setzer, 1971; Rosevear, 1974; van Gelder, 1978). Rüppell's foxes are physically similar to *V. bengalensis* and the other desert fox species (*V. chama, V. velox, V. corsac, Fennecus zerda*) (Clutton-Brock *et al.*, 1976). There are five or six subspecies (Rosevear, 1974; Meester and Setzer, 1971; Stains, 1975). Rosevear (1974) remarked that there is little consistent physical variation between specimens from various areas, although those from northeast Africa are slightly smaller.

DIET

Rüppell's foxes are omnivorous. Small mammals constitute their primary food source. They also eat lizards, snakes, birds, insects, berries, and plant roots (Lindsay and Macdonald, 1986; Kowalski, 1988; Meester and Setzer, 1971; Dorst and Dandelot, 1969). Scat analysis from a population in Oman indicated a primary reliance on small mammals. Lizards and insects were important dietary elements, and grass and scavenged material were also present in relatively small amounts (Lindsay and Macdonald, 1986). In Pakistan Rüppell's foxes are found near extensive rodent colonies and probably feed on them (Roberts, 1977). In the Egyptian Sahara, scat analysis showed an omnivorous diet, including gerbils, birds, dates, insects, snakes, and human refuse (Kowalski, 1988). In Saudi Arabia the foxes travel along wadis where there are abundant rodent population (Harrison, 1968). Rüppell's foxes also frequent garbage dumps and eat scraps scavenged from areas of human habitation (Lindsay and Macdonald, 1986; Harrison, 1968; Kowalski, 1988). A captive ate meat, fruit, vegetable greens, and biscuits, and also killed and ate live prey. It never drank water (Petter, 1952). This captive would fill its mouth with large quantities of food which it would carry to its bedding area, then repeat the process one or two times before finally settling down to eat (Petter, 1952). These foxes can survive with extremely limited access to water (Kowalski, 1988).

ACTIVITY

Free-ranging foxes in Oman rested underground in their dens during the daytime, and emerged at night (Lindsay and Macdonald, 1986).

REPRODUCTION

Litter size is probably two or three pups (Roberts, 1977; Rosevear, 1974). Roberts (1977) mentioned that a litter dug out of a sandhill in the Mekran coastal area was probably born in early March. Petter (1952) reported that a captive held in France showed testicular regression in the spring, with the testes fully developed in October. Mated pairs in Oman often denned together, and males denned close to their mates while the cubs were young (Lindsay and Macdonald, 1986). A captive in a London zoo lived 6.5 years (Flower, n.d., cited in Shortridge, 1934).

SOCIAL ORGANIZATION AND BEHAVIOR

Rüppell's foxes live in mated pairs. Groups of three to five have been observed: It is likely that these are family groups (Meester and Setzer, 1971; Roberts, 1977; Dorst and Dandelot, 1969). Rüppell's foxes are not as gregarious as pale foxes (*V. pallida*) (Rosevear, 1974).

Lindsay and Macdonald (1986) studied six radio-collared foxes in Oman. Each lived as a member of a mated pair. Each pair sometimes shared a den, either in turns or sometimes simultaneously. Home ranges of these pairs were essentially the same, and there was no overlap between the home ranges of neighboring pairs. The mean size of the home ranges for the six Rüppell's foxes was 69.1 km², with males' ranges slightly larger than those of females. A track ran along the boundary between two of these home ranges, and foxes walked up and down this track, barking. All observed barking occurred at home range peripheries. Den sites were scattered throughout the foxes' home ranges. Foxes frequently moved from den to den (average 4–7 days, range 1–21 days). Two out of three pairs had cubs (Lindsay and Macdonald, 1986).

Petter (1952) described the behavior of a captive Rüppell's fox raised by humans from the age of 3 months. This male fox became very tame and frequently exhibited a variety of play behaviors. Vocalizations of Rüppell's foxes include barks, directed toward conspecifics at territorial boundaries (Lindsay and Macdonald, 1986); growls, emitted by a captive, directed toward unfamiliar individuals (Petter, 1952); chattering and murmuring sounds that function in greeting contexts (Petter, 1952); and sonorous yelps, which have no known function but perhaps are related to contact or breeding (Petter, 1952).

Vulpes velox: Swift Fox

Swift foxes are one of the three species in the genus *Vulpes* that exist in North America. The common names "swift fox" and "kit fox" have been used interchangeably for *Vulpes velox* and *Vulpes macrotis*, although the former name seems preferable for *Vulpes velox* since it is a translation of the Linnaean name. Some authorities consider swift and kit foxes as a single species (see *taxonomy* section below). From the nineteenth through the mid-twentieth century, human activities brought about a widespread reduction in swift fox numbers. Predator-poisoning programs, habitat destruction, hunting, and trapping all contributed to this decline. In the last 25 years swift foxes seem to be returning to much of their historic range, and they are now present in many areas where they were once extirpated.

DISTRIBUTION AND HABITAT

The original range of swift foxes included much of the plains of west-central North America from the Texas panhandle northward into the prairie provinces of southern Canada (Egoscue, 1979; Banfield, 1974; Muchmore, 1975). By the mid 1900s, after a prolonged period of population reduction due to human activities, the species range had shrunk to a small central area. Since the 1950s swift foxes have begun to reoccupy parts of their original range, and they have reappeared in Oklahoma, Texas, Kansas, Nebraska, Wyoming, and Montana (Moore and Martin, 1980; Laurion, 1988; Floyd and Stromberg, 1987; Zumbaugh and Choate, 1985; Clark and Stromberg, 1987; Samuel and Nelson, 1982). See Scott-Brown *et al.* (1987) for a detailed distribution map. In 1978 the Committee on the Status of Endangered Wildlife in Canada classified the swift fox as extirpated (in Canada). Beginning in 1983 swift foxes were released at a number of sites in Alberta and Saskatchewan. Releases have continued in these two provinces up to the present (Carbyn, 1986, 1989b; Lynch, 1987), and as of 1990 some 246 foxes have been returned to the wild (Ginsberg and Macdonald, 1990). Predation by coyotes (*Canis latrans*) and human-caused deaths have been significant causes of mortality at the reintroduction sites (Carbyn, 1989b). It is not yet evident whether the reintroduced individuals can establish viable, self-sustaining populations.

Swift foxes are animals of the open plains, inhabiting both short and medium grass prairies. They require burrow and den sites for year-round use (unlike most other canid species whose use of dens is seasonal). These dens may be dug by the foxes themselves or modified from the existing excavations of prairie dogs, badgers, or other animals. Some dens, particularly those used for whelping, are large and complex with many entrances. They are located on open plains, hilltops, and other well-drained sites (Uresk and Sharps, 1986; Egoscue, 1979; Laurion, 1988; Kilgore, 1969; Cutter, 1958a; Floyd and Stromberg, 1981).

PHYSICAL CHARACTERISTICS

Swift foxes are small and slight, not much larger than house cats. They are markedly smaller than red foxes (*V. vulpes*) and only slightly larger than kit foxes (*V. macrotis*). Males are slightly heavier (averaging 2–2.4 kg) and larger than females (averaging 1.9–2.2 kg). Weights for either sex can range from 1.6 to almost 3 kg. Body length including head plus tail is 60–84.4 cm, with males averaging slightly longer than females. Tail length is 22.5–28 cm (Kilgore, 1969; Egoscue, 1979; Banfield,

1974; Hall, 1981; Stains, 1975). The young attain adult size by August of their first year. Pelage color varies from buff-yellow to buff-gray to grizzled gray. The underfur is thick, and guard hairs are scattered sparsely. In winter the coat is long and dense, but after the spring molt in April to July the fur is short, thin, and harsh with more of a red-gray tone to it. The flanks and legs may be orange-tan to ochraceous buff, and the feet are no darker than the rest of the body. The foot pads are almost completely covered by coarse fur. The legs are proportionately shorter than in *V. vulpes*. The underparts, throat, chest, and belly are pale, ranging from cream to pure white (Egoscue, 1979; Banfield, 1974; Wendt, 1975a; Hall, 1981; Muchmore, 1975). The well-furred tail is grizzled gray above, orange-tan below, with a black tip. There is a slight black patch over the tail gland. The head is grizzled gray on the forehead, ochraceous buff on the back of the ears, and the cheeks are whitish. The insides of the ears are cream to white. There are two black or brown-black spots on the face, one on either side of the muzzle below the eyes. The snout is broader than in *V. macrotis* and the ears are shorter (Egoscue, 1979; Muchmore, 1975; Banfield, 1974; Stains, 1975; Wendt, 1975a). The northern subspecies, *V. v. hebes*, is slightly larger and paler with more of a gray color to the pelage than *V. v. velox*, the southern subspecies (Muchmore, 1975).

The skull is smaller and lighter than that of *V. vulpes*, although the auditory bullae are relatively larger (Banfield, 1974). The bullae are smaller than in *V. macrotis*, however, and the rostra are wider in swift foxes (Dragoo *et al.*, 1987; Stains, 1975). Clutton-Brock *et al.* (1976, p. 159) described the skull as "typically `fox-like'." They stated that it is quite similar to that of *V. chama* and *V. bengalensis*, although the upper molars are slightly less developed than in the latter species. See Egoscue (1979) for cranial measurements. The dental formula is incisors 3/3, canines 1/1, premolars 4/4, molars 2/3 = 42.

TAXONOMY

A number of sources consider swift foxes (*V. velox*) and kit foxes (*V. macrotis*) as a single species (Dragoo *et al.*, 1990; Hall, 1981; Clutton-Brock *et al.*, 1976; van Gelder, 1978; Ginsberg and Macdonald, 1990; Ewer, 1973). Marked differences in morphological characteristics exist (Dragoo *et al.*, 1987, 1990), and electrophoretograms of serum protein and hemoglobin fractions are recognizably different (Thornton and

Creel, 1975). The two species interbreed where their ranges coincide, but the region of hybridization is quite restricted, and the evidence "suggests that selection generally opposes hybrids and favors maintenance of separate adaptive modes" (Rohwer and Kilgore, 1973, p. 163). So despite this occasional hybridization, the species are readily differentiable. In conflict with this conclusion, however, recent biochemical analyses show a lack of genic differentiation between the two species, indicating that perhaps only a subspecific level of separation is warranted (Dragoo *et al.*, 1990). Dragoo *et al.* (1990) therefore suggested that *V. macrotis* and *V. velox* are conspecific, the name *V. velox* (Say) having priority, and that all heretofore recognized subspecies of both should fall into one or the other subspecies.

A revision of taxonomy at the subspecific level is now overdue. Two subspecies of *V. velox* have been recognized: *V. v. velox* from the southern part of the range and *V. v. hebes* from the northern. Recent analysis of the differences between groups of specimens from these subspecies (and a third group of specimens from Montana) indicates that subspecific classification is "probably not justified" (Stromberg and Boyce, 1986, p. 105). Although significant geographic variation exists, it may be clinal. Furthermore, there are no geographic barriers to gene flow. Although it has been delisted by the U.S. government, *V. v. hebes* is listed as endangered by the IUCN (Ginsberg and Macdonald, 1990).

DIET

Swift foxes are opportunistic feeders. Their diet varies seasonally and with the availability of food. Mammals the size of black-tailed jackrabbits and smaller are the principal food everywhere. Other mammals consumed include cottontail rabbits, ground squirrels, pocket mice, kangaroo mice, and various other small rodents (Uresk and Sharps, 1986; Zumbaugh *et al.*, 1985; Kilgore, 1969; Cutter, 1958b; Loy and Fitzgerald, 1980; Egoscue, 1979). Flocking and ground-nesting birds are important food sources; in Oklahoma they are the second most important dietary component (Kilgore, 1969). Swift foxes are adept at catching prairie chickens, but generally do not take other game birds and domestic poultry (Kilgore, 1969; Seton, 1937, cited in Egoscue, 1979; Loy and Fitzgerald, 1980; Cutter, 1958b). They frequently consume insects, most often Orthopterans and Coleopterans. In Texas, insects (primarily grasshoppers and beetles) made up 29% of stomach contents and 55% of scats (by volume) (Cutter, 1958b). The relative importance of insects to the diet varies seasonally (Kilgore, 1969). Various reptiles and amphibians are eaten. Lizards are an important feature of the diet in Texas. Fish trapped in evaporating playas are caught and eaten. Although some may be ingested accidentally, grasses and berries provide a significant source of nutrients (Cutter, 1958b; Banfield, 1974; Egoscue, 1979; Kilgore, 1969). Carrion is sometimes an important food source, particularly in winter months (Scott-Brown et al., 1987; Zumbaugh et al., 1985). Average food consumption per fox per day is 200 g (Kilgore, 1969).

Swift foxes sometimes cache food (Banfield, 1974). Bunker (1940) observed that members of this species were always first to take poison placed to kill wolves at buffalo carcasses, a habit that contributed to their decline.

ACTIVITY

Swift foxes are essentially nocturnal. They usually spend daylight hours asleep in their burrows. Adults and pups may be seen aboveground in the daytime, but they always remain close to the den, and their activities are restricted to sunning and resting (Kilgore, 1969; Cutter, 1958a; Muchmore, 1975; Banfield, 1974; Egoscue, 1979). Some diurnal foraging, probably minimal, is suggested by the presence of diurnal prey species in scats (Kilgore, 1969).

REPRODUCTION

Females are annually monestrous. They may breed in their first year (Scott-Brown et al., 1987). Breeding occurs from late December to early February. Gestation is approximately 51 days (Scott-Brown et al., 1987). Pups are born in March, April, or May. Litter sizes average four or five (range one to six). By August pups are full grown, and they remain with their parents until autumn (Laurion, 1988; Kilgore, 1969; Egoscue, 1979; Banfield, 1974; Loy and Fitzgerald, 1980). Adults live in pairs; they may mate for life (Jones et al., 1985). Occasionally one male and two females share a single burrow, an arrangement that also occurs in V. vulpes and V. macrotis (Banfield, 1974; Egoscue, 1979). Males assist in provisioning the young (Seton, 1909, cited in Egoscue,

1979). Seton (1909, cited in Egoscue, 1979) observed both parents of a pair attempting to decoy a dog away from their burrow where the pups were. Swift foxes are easy to maintain in captivity, but they will seldom rear young in exhibit cages unless a secure den is provided (Egoscue, 1979).

Swift foxes are preyed upon by a number of species, but the most catastrophic mortality has been due to humans. Poisoning campaigns directed against coyotes or wolves, automobiles, hunting, and trapping all contribute to mortality. Members of this species are relatively easily trapped. They often den near human habitation, which adds to their vulnerability. Coyotes are a major cause of mortality in some areas (Laurion, 1988). Longevity in the wild is unknown, but is probably less than 7 or 8 years. Captives have lived as long as 12 years 9 months (Crandall, 1964, cited in Egoscue, 1979).

SOCIAL ORGANIZATION AND BEHAVIOR

The essential social unit is the mated pair. Occasionally, groups of one male and two females occur (Banfield, 1974; Egoscue, 1979). The relationship between the females in these trios has not been studied; perhaps one is a helper or subordinate. The young of the year remain with their parents until autumn when they disperse. Nothing is known about dispersal patterns.

In Colorado, mark-recapture data indicated home ranges of 172–210 ha for three males and 86 ha for one female (Fitzgerald *et al.*, 1981, cited in Scott-Brown *et al.*, 1987). In Nebraska, home ranges vary in size from 10 to 44 km^2, the average being approximately 25 km^2 (Laurion, 1988). In Colorado there is extensive home range overlap (Laurion, 1988). Banfield (1974) gave the population density on ideal range as one fox per 5 km^2. Nothing is known about territoriality in this species. Vocalizations include purrs, growls, whines, and shrill yaps (Banfield, 1974).

Vulpes vulpes: Red Fox

Red foxes have the largest range of any species in the family Canidae [though before intense eradication efforts by man in the last century the natural range of gray wolves (*Canis lupus*) was larger]. In fact, red foxes have the largest natural distribution of any living terrestrial mammal except man (Nowak and Paradiso, 1983). Unlike wolves, whose social structure cannot withstand even moderate pressure from human hunting or habitat destruction, red foxes maintain and some-times increase their numbers in the face of these pressures. Bounties have been ineffective in reducing populations in several midwestern and New England states (Ables, 1975). This is a commercially impor-tant species. In addition to the huge numbers raised on fur farms, many are trapped for fur each year. Close to half a million were killed in the 1976–77 season in the United States and Canada (Nowak and Paradiso, 1983). Prime pelts go for as much as 50 dollars (Jones and Bir-ney, 1988). Red foxes are hunted for sport as well.

DISTRIBUTION AND HABITAT

Present range encompasses most of the Northern Hemisphere north of 30 degrees north latitude (Hersteinsson and Macdonald, 1982). The whole of Asia except its extreme southeastern portion is included, as are Europe, North Africa, and North America south to central Texas. Red foxes were introduced to Australia and some Pacific islands in the 1800s, where they have flourished. Some individuals from England were introduced to northeastern North America between 1650 and 1750, and although there was already an indigenous population, the in-troduced animals thrived. After this introduction, although not neces-sarily because of it, red foxes increased their range to include the southeastern United States and spread westward to the Great Plains (Lloyd, 1975, 1980; Stains, 1975; Zimen, 1980; Hersteinsson and Mac-donald, 1982). Northern limits of distribution are generally imposed by tundra, although foxes range north to the Arctic Ocean in some areas. They are known to travel on the sea ice offshore from the Labrador coast (Andriashek et al., 1985).

Red foxes are found in an enormous variety of habitats, including forests, mixed woodland, meadows, plains, river valleys, and moun-tainous regions (Lloyd, 1980; Ables, 1975; Voigt, 1987; Novikov, 1962). Although they generally avoid arid regions (Furley, 1986), in northwest India some red foxes live among sand dunes and in scrub-covered wasteland (Prater, 1965). Agricultural land is often prime fox habitat (Hewson, 1986), and some of the highest densities in North America are in heavily populated farm and dairy communities (Ables, 1975). Village and urban outskirts are favored as well (Novikov, 1962), and red foxes live within the urban areas of some large cities (Zimen, 1980; Doncaster *et al.*, 1990). In mountainous regions their range extends above timberline. In the alpine regions of the inner Himalayas red foxes are found up to 4,500 m (Mitchell, 1977; Haltenorth and Roth, 1968; Novikov, 1962). In general, broken or diverse biotopes are favorable (i.e., forest interspersed with meadows). In England "diversity and fragmentation of the habitat within small areas" supports more numerous populations (Lloyd, 1975, p. 207). This species is rare or absent in unbroken forest, dense lowlands, pine forests of the southeastern United States semi-arid grasslands of North America, and deserts (Lloyd, 1975; Novikov, 1962).

Coyotes (*Canis latrans*) influence the distribution of red foxes in some regions. Coyotes are often "overtly antagonistic" toward red foxes, sometimes killing them and their cubs (Sargeant and Allen, 1989, p. 632). Red fox distribution in Alberta is confined to areas without coyotes (Dekker, 1983). Available habitat in Maine and North Dakota is strictly limited by coyote territories as well (Harrison *et al.*, 1989; Sargeant *et al.*, 1987). In other regions there is a "high degree of interspecific tolerance" (Sargeant and Allen, 1989, p. 632). In the southwest Yukon red foxes and coyotes are sympatric and rely on the same prey species, although there are some differences in habitat preference (Theberge and Wedeles, 1989).

Most of the foxes' life is spent aboveground, although parturition and cub rearing occur in dens. Dens may be dug by the foxes themselves or modified from the existing excavations of other animals (Lloyd, 1980; Ables, 1975). Some dens have been occupied for at least 35 years (Ables, 1975), and these may be quite elaborate with as many as 19 entrances. Each family group has a primary den and also uses auxiliary burrows or takes shelter among rocks in an emergency.

PHYSICAL CHARACTERISTICS

V. vulpes is the largest of the ten species in the *Vulpes* genus. Head-plus-body length ranges from 60 to 90 cm, height at shoulder is 30–40 cm and tails are 30–60 cm long (Haltenorth and Roth, 1968; Novikov, 1962). Weights range from 3 to 14 kg: individuals over 10 kg are rare

(Zimen, 1980; Haltenorth and Roth, 1968). Males are consistently larger and heavier than females. Ables (1975) gave the average weight range of males as 4.5–5.4 kg, that of females as 4.1–4.5 kg, and indicated that in Canada and northern areas weights are greater. For foxes in the U.S.S.R. Novikov gave male weight ranges of 6–10 kg, females 5–8 kg.

The common name of red fox is to some degree a misnomer since by no means all of these foxes are rufous. Typical pelage colors range from pale yellow-red through tawny, red-brown, red-black, and brownish-black, passing through an infinity of transitional shades (Haltenorth and Roth, 1968; Stroganov, 1962; Nowak and Paradiso, 1983; Novikov, 1962). Several naturally occurring color phases and melanistic forms are common. Silver foxes are predominantly black, with silver-tipped guard hairs, and cross foxes are distinguished by a dark marking formed by the convergence of dorsal and transverse saddle marks. Entirely black individuals occur (Haltenorth and Roth, 1968; Ables, 1975; Voigt, 1987; Prater, 1965; Clutton-Brock et al., 1976). All three color phases (silver, cross, and red) may occur in the same litter (Murie, 1944, cited in Ables, 1975). Partial and complete albinos occur, but only very rarely (Novikov, 1962). Chest and underparts are white to gray, and the undercoat is often gray (Novikov, 1962; Nowak and Paradiso, 1983). On the whole, cross foxes make up 25% and silver foxes 10% of the species (Nowak and Paradiso, 1983). Breeders have successfully selected for a variety of other true-breeding pelage shades, including white, snowy, multicolored, platinum, and white-muzzled (Novikov, 1962). It is worth noting that Keeler (1975) described behavioral traits that consistently attach to various color phases.

The head is narrow with a long, pointy muzzle. The ears are pointed and black or dark brown on the back. The insides of the ears and the chin are light in color. There may be a dark marking between the nose and eyes. The pupils are elliptical, not round as they are in *Canis*. Limbs are black or dark brown on their lower parts and anteriorly (Clutton-Brock et al., 1976; Nowak and Paradiso, 1983; Novikov, 1962; Stroganov, 1962). The skull is slender and elongate, the braincase roughly equal in length to the facial region, which is itself long and narrow. The canines are long and finely pointed, the carnassials rather small and very sharp. The dental formula conforms to the usual canid pattern of incisors 3/3, canines 1/1, premolars 4/4, molars 2/3 = 42 (Clutton-Brock et al., 1976; Novikov, 1962; Stroganov, 1962).

TAXONOMY

A thorough overhauling of the taxonomy of the *Vulpes* genus is over-due, as is revision of the 40-odd subspecies described for *V. vulpes* (Zimen, 1980). In the past a number of authorities conferred specific status on the North American red fox (*V. fulva*). More recently *V. fulva* has been considered conspecific with the Palearctic *V. vulpes* (Voigt, 1987). Forty-eight subspecies have been described for *V. vulpes*, "giving this canid the largest taxonomic complex" of all the canids (Stains, 1975, p. 18). Twelve subspecies have been described in North America alone (Ables, 1975).

Van Gelder (1978) suggested that the entire genus *Vulpes* be included as a subgenus of *Canis*, and that the red fox be recognized as *Canis (Vulpes) vulpes*. This reorganization is unlikely to be accepted, if only because usage of the *Vulpes* genus is so deeply entrenched. Furthermore, recent genetic analyses suggest that this revision would be inappropriate, since *Vulpes* is quite distinct from *Canis* (Simonsen, 1976; Wayne and O'Brien, 1987).

DIET

Over 100 dietary studies of red foxes have been conducted (Hersteinsson and Macdonald, 1982), and most found that small mammals, primarily rodents, constituted the primary food source. But overall, red foxes are paragons of dietary opportunism. Besides being predators, red foxes are also scavengers, insectivores, and frugivores. Season, habitat, and availability all influence dietary patterns. Dietary flexibility and opportunism enable red foxes to exist where more specialized species could not. In an English population, Macdonald (1977, p. iii) found diet to vary enormously between habitats and even between neighboring territories: "The pattern of availability of food, as distinct from its abundance, may indirectly determine group size, heterogeneously distributed prey favoring larger groups." Lloyd (1980) has suggested that some individuals have specialized feeding patterns. This suggestion is compatible with consumption patterns in other dietary generalist species. Foxes eat rodents, including moles, voles, mice, pocket gophers, red squirrels, and hares; insects, including grasshoppers, bumblebees and their honey, beetles, wasps, butterflies, flies and their larvae; and also frogs, toads, snails, and lizards. They eat earthworms, often as a staple food. Birds, particularly passerines and waterfowl and their eggs, can be important prey. Red foxes are often considered a threat to poultry, although depredations are generally localized. Larger prey include young chamois, young roe deer, lamb, kids, domestic cats, marmots, martens, and young boar. The upper size limit of prey seems to be about 3 kg. Carrion, including offal and afterbirths, is important in the diet in some regions, as is human refuse of every description. Red foxes consume an enormous variety of vegetable material. During autumn, fruit may make up 100% of food intake. Berries (blueberries, blackberries, raspberries), cherries, apples, plums, tubers including potatoes and turnips, grains such as oats and maize, cabbage, acorns, fresh grass, and moss are all eaten. (Henry, 1986; Hersteinsson and Macdonald, 1982; Doncaster *et al.*, 1990; Furley, 1986; Dekker, 1983; Lloyd, 1975; Ables, 1975; Haltenorth and Roth, 1968; Novikov, 1962; Nowak and Paradiso, 1983). In littoral habitats fish, crayfish, marine birds, their eggs and fledglings, and all sorts of detritus are consumed (Haltenorth and Roth, 1968). Along the arctic coast red foxes scavenge on ringed seal carcasses (Andriashek *et al.*, 1985). On average in central Europe, red foxes consume 0.5–1 kg of food per day per adult (Haltenorth and Roth, 1968).

The habit of caching food is universal among red fox populations, although excess may be carried back to the den as well (Henry, 1986; Dekker, 1983). Cached items may be scatter-hoarded or hoarded in groups (Hersteinsson and Macdonald, 1982).

ACTIVITY

Red foxes are primarily nocturnal, or nocturnal with crepuscular peaks of activity. During mating season and when the kits are young, diurnal activity is common as well. Daytime activity generally increases in winter (Henry, 1986; Ables, 1975; Haltenorth and Roth, 1968; Roberts, 1977; Voigt, 1987; Novikov, 1962). Haltenorth and Roth (1968) found three activity peaks: 0400–0500 hours, 1900–2000 hours, and 0000–0100 hours.

REPRODUCTION

Females are annually monestrous, and mating occurs in late winter to early spring (December to April), depending on latitude (Hersteinsson and Macdonald, 1982). In North America breeding peaks occur from late December to early February (Ables, 1975). During breeding season vixens may be attended by several males who participate in ritualized fights. After a gestation period of 49–56 days (usually 51–52 days) from 1 to 13 cubs are born. The usual litter size is five pups (Jones and Birney, 1988). Ten weeks after parturition, after litters have emerged from the den and been weaned, there are far fewer left alive. Lloyd (1975) found litter size at this time to average 2.7 (Haltenorth and Roth, 1968; Dekker, 1983; Henry, 1986; Banfield, 1974; Lloyd, 1975; Novikov, 1962; Ables, 1975).

Reproductive behaviors are far more complex than the conventionally described short-term pair bond. A number of reproductive strategies exist, including monogamy, polygyny, and temporary nonbreeding. Observations of captives on fur farms show that some pairs are monogamous but that polygyny is common (Tembrock, 1957, cited in Haltenorth and Roth, 1968; Novikov, 1962). In the wild, social groups of a male and more than one vixen are by no means rare. Within these groups, characteristically only the dominant vixen reproduces (Macdonald, 1977; Lindstrom, 1989). Thus in the larger social groups only a minority of vixens rear young, and nonbreeding females may constitute a large fraction of a population. Lloyd (1975) found the proportion of nonreproductive females averaged over four years to be 14–29% of the total population. Nonbreeders that are part of a family group may provide alloparental care. In some cases, they feed, guard, sleep with, and generally behave "amicably" toward another female's cubs (Macdonald, 1980, p. 133). Von Schantz (1981) observed an instance of a free-ranging, nonbreeding female raising a litter of cubs whose mother had died. This certainly is an instance of significant alloparental contribution to pup care. The significance of the participation of other nonbreeding females to the cubs themselves is unknown. Whether the cubs actually benefit remains to be determined, since alloparental care does not necessarily contribute significantly to the survival rate of the recipients (Moehlman, 1983). Kinship groups may simply result from resource surpluses that permit the nonbreeders to delay dispersal (von Schantz, 1984). The mechanism of reproductive suppression in sexually mature, nonbreeding females is unknown. In a radio-tracking study of free-ranging individuals in Sweden, von Schantz (1981) determined that nonbreeding females regularly became pregnant but aborted or deserted their young.

On some occasions more than two litters may be born within a group, and these litters may be communally reared in one den. Macdonald (1980) reported that a "mixed litter," the offspring of two different females, was attended and suckled indiscriminately by both. He records that he has "made sufficient observations of multiple litters in England to believe that they are a widespread phenomenon" (Macdonald, 1980, p. 128). Males of mated pairs participate in parental care, provisioning the postparturient female and young. Males may cache food around the natal den. They remain close to their mates during cub rearing and play with their offspring (Macdonald, 1977, 1979a, 1980; Ables, 1975; Haltenorth and Roth, 1968). Family groups composed of a mated pair and their offspring (and in some cases other affiliated females) remain together in their home range until autumn. During autumn and winter the young disperse and establish territories in new areas, or perish (Lloyd, 1975; Ables, 1975). A field study of dispersal in Bristol, England, showed significantly diminished survival in dispersing (versus nondispersing) foxes, indicating that there may be very large risks associated with dispersal (Woollard and Harris, 1990).

In many habitats mortality is high and population turnover great; therefore, "range occupancy is in a state of flux throughout much of the year" (Lloyd, 1980, p. 20), and potential range frequently becomes available. Most subadult males emigrate from their natal home range within a year. Females emigrate less frequently (Hersteinsson and Macdonald, 1982; Lindstrom, 1989; von Schantz, 1981; Macdonald, 1979a; Storm and Montgomery, 1975; Lloyd, 1975). Some vixens will not disperse, instead remaining behind and becoming incorporated into the natal group (Hersteinsson and Macdonald, 1982). Dispersed individuals settle and breed at 10–12 months of age (Lloyd, 1980). The distances traveled by dispersing individuals "seem to be related to fox population density or to some parameter related to it" (Lloyd, 1975, p. 212). Young foxes may disperse distances of a few km to over 200 km. In general, males disperse greater distances than females (Hersteinsson and Macdonald, 1982; Ables, 1975). A red fox population in south-central Sweden appears to be limited by food levels in years of low vole densities, and by social regulation during years of increasing and high vole densities (Lindstrom, 1989), indicating that population regulation may be subject to complexes of controlling factors.

Red foxes suffer high mortality rates. Predominant causes of death include parasites and disease. Rabies is responsible for extensive mortality, and red foxes are a primary vector in many regions (see Macdonald and Voigt, 1985). Predation by coyotes, dingoes, domestic dogs, bear, lynx, wolves, and golden eagles, and human activities such as shooting, trapping, and roadkills, are also regionally important (Dekker, 1983; Voigt, 1987; Lloyd, 1980; Haltenorth and Roth, 1968). Potential longevity is on the order of 12 years, although the average natural lifespan of free-ranging individuals is less than 1 year, and few live beyond 3–4.5 years (Roberts, 1977; Ables, 1975; Nowak and Paradiso, 1983).

SOCIAL ORGANIZATION AND BEHAVIOR

Recent studies of sociality in red foxes have disclosed a far greater degree of complexity and flexibility than was previously thought to exist. In some regions the social unit is the mated pair, while in others more complex social groupings occur, involving one adult male and several adult females. Rarely, social groups involving more than one male may exist. Some foxes are itinerants without a home area (Macdonald, 1977). Everywhere the young of the year are integrated into the social fabric of the family group from summer into the fall or winter when most of them disperse. Some females remain in their natal home range and do not disperse at all. The majority of observations on the larger social groups have been made by Macdonald (1977, 1979a, 1980). In England he found that

> groups typically comprised one dog-fox and a variable number of related vixens. A detailed study of the social behaviour within and between family groups indicates a highly integrated society within which the adult vixens of a group form a social hierarchy which affects access to food and the probability of reproducing. Within larger groups, only the dominant vixen reproduces. Subordinate vixens contribute actively to the well-being of the dominant's cubs (to whom they are closely related).

"All available evidence" suggests that where groups of several vixens occur, they are close relatives (Macdonald, 1980, p. 123). In some cases the two females of a group are mother and daughter (Niewold, 1980). Vixens have their own dominance hierarchy within their group, and all are generally subordinate to the dog fox (Hersteinsson and Macdonald, 1982). Lloyd (1980) suggested that complex social groups may be characteristic of populations "unaffected adversely by man." "Field evidence collected in Wales and North America strongly suggests that most foxes exist as pairs" (Lloyd, 1980, p. 21). Some authors describe an even less social existence. According to Ables (1975, p. 233), North American red foxes remain "more or less solitary" between the time they disperse and the following mating season.

Typically, red foxes are territorial, with family groups occupying and defending a given area. The flexibility apparent in other facets of red foxes' social organization is present here as well, and "less clear cut spatial systems" also exist (Hersteinsson and Macdonald, 1982, p. 266). Niewold (1980) pointed out the dynamic nature of the territorial structure, which is not at all a fixed system. Dispersing individuals seem to find and occupy a home range and, once established there, usually remain for life (Ables, 1975). In North American populations, activities are restricted to home ranges from spring through early fall (the kit-rearing season). Neighboring family groups occupy well-defined, nonoverlapping home ranges and generally avoid one another (Storm and Montgomery, 1975; Goszczynski, 1989). The social groups observed by Macdonald also occupied exclusive territories from which neighboring groups were excluded. Not all boundaries were clear, however. Some territorial limits were clearly defined and defended, while in other places they were "hazy," and territories were not strictly exclusive (Macdonald, 1981). On the whole, there is little overlap between home ranges (Ables, 1975). Foxes intruding into adjacent territories are sometimes attacked and expelled; the attacks are sometimes intense (Hersteinsson and Macdonald, 1982). In the Netherlands Niewold (1980) observed many aggressive interactions between red foxes.

Both dogs and vixens scent-mark using urine and feces (Burrows, 1968). Urine marking is thought to serve a territorial marking function (Haltenorth and Roth, 1968; Macdonald, 1977), and neighbors seem to recognize one another's marks. Barash (1974) found a significantly higher incidence of dominant-subordinate relationships between neighboring foxes (who had presumably had opportunities to establish this relationship) than between strangers who had not encountered one another before.

Home range sizes are highly variable. They are determined at least in part by resource availability and distribution and are influenced by terrain and habitat complexity as well (Macdonald, 1977; Ables, 1975; Goszczynski, 1989). Home range size varies from 0.1 to 20 km² (Hersteinsson and Macdonald, 1982). In normal, varied habitat, home range size is 2–5 km² (Haltenorth and Roth, 1968). Population density is highly variable. Haltenorth and Roth (1968) found an average of one fox per ha. In prime habitat one square kilometer may support one or two adults (Nowak and Paradiso, 1983). Von Schantz (1981) found that nonbreeding females had smaller, suboptimal home ranges within the larger home ranges of breeding females. Storm (1965, cited in Ables, 1975) found that home ranges consist of areas of intense use connected by pathways. Ranges may expand during the winter (Sheldon, 1949). Itinerant foxes who have not established themselves in a territory do not have a good chance of long-term survival (Macdonald, 1980).

Red fox vocalizations include a variety of agonistic and affiliative sounds. Whines and whimpers occur in contexts of distress or submission. Mew calls are short-range affiliative vocalizations (Cohen and Fox, 1976). Growls occur in contexts of defense and threat. Barks are threat and territorial defense sounds. Screams are emitted during intraspecific dominance conflicts, and in contexts of defense and intense distress (Cohen and Fox, 1976; Barash, 1974). Coos, observed only in red foxes, are greeting and contact-solicitation calls (Cohen and Fox, 1976; Marten, 1980). There is no group long-range vocalization or howl.

BIBLIOGRAPHY

Ables, E. C. 1975. "Ecology of the red fox in North America." In *The Wild Canids* (M. W. Fox, ed.), pp. 216–236. New York: Van Nostrand Reinhold.

Acharyjo, L. N. and R. Misra. 1976. "A note on the breeding of the Indian fox (*Vulpes bengalensis*) in captivity." *J. Bombay Nat. Hist. Soc.* 73:208.

Acosta, A. 1972. "Hand rearing a litter of maned wolves *Chrysocyon brachyurus* at Los Angeles Zoo." *Int. Zoo Yearb.* 12:170–174.

Afik, D. and P. U. Alkon. 1983. "Movements of a radio-collared wolf (*Canis lupus pallipes*) in the Negev Highlands, Israel." *Isr. J. Zool.* 32:138–146.

Alcorn, J. R. 1946. "On the decoying of coyotes." *J. Mammal.* 27(2):122–126.

Allen, D. L. 1979. *The Wolves of Minong.* Boston: Houghton Mifflin.

Allen, G. M. 1923. "The Pampa fox of the Bogata savanna." *Proc. Biol. Soc., Washington* 36:55–58.

Allen, S. H., J. O. Hastings, and S. C. Kohn. 1987. "Composition and stability of coyote families and territories in North Dakota." *Prairie Nat.* 19(2):107–114.

Altmann, D. and W. Recker 1971. "Verhaltensanalyse der Ontogenese von Steppenfüchsen, *Vulpes corsac* L." *Zool. Gart., N. F., Leipzig* 41(1/2):1–6.

Andelt, W. F. 1982. *Behavioral ecology of coyotes on the Welder Wildlife Refuge, south Texas.* Ph.D Diss., Colorado State University, Fort Collins.

Andelt, W. F. 1985. "Behavioral ecology of coyotes in south Texas." *Wildl. Monogr.* 94: 1–45.

Andelt, W. F. and B. R. Mahan. 1980. "Behavior of an urban coyote." *Am. Midl. Nat.* 103(2):399–400.

Andelt, W. F., J. G. Kie, F. F. Knowlton, and K. Cardwell. 1987. "Variation in coyote diets associated with season and successional changes in vegetation." *J. Wildl. Manage.* 51(2):273–277.

Anderson, S. and J. K. Jones, Jr. (eds.). 1984. *Orders and Families of Recent Mammals of the World.* New York: John Wiley and Sons.

Andriashek, D., H. P. L. Killian, and M. K. Taylor. 1985. "Observations on foxes, *Alopex lagopus*, and *Vulpes vulpes*, and wolves, *Canis lupus*, on the off-shore sea ice of northern Labrador." *Can. Field-Nat.* 99(1):86–89.

Artois, M. and M.-J. Duchêne. 1982. "Première identification du chien viverrin (*Nyctereutes procyonoides* Gray, 1834) en France." *Mammalia* 46(2):265–267.

Asa, C. S. and M. P. Wallace. 1990. "Diet and activity pattern of the Sechuran desert fox (*Dusicyon sechurae*)." *J. Mammal.* 71(1):69–72.

Asakura, S. 1968. "A note on the breeding records of Japanese foxes *Vulpes vulpes japonica* at Tokyo Tama Zoo." *Int. Zoo Yearb.* 8:20–21.

Avery, G. and D. M. Avery. 1987. "Prey of coastal black-backed jackal *Canis mesomelas* (Mammalia: Canidae) in the Skeleton Coast Park, Namibia." *J. Zool., London* 213:81–94.

Ballard, W. B. and J. R. Dau. 1983. "Characteristics of Gray wolf, *Canis lupus,* den and rendezvous sites in southcentral Alaska." *Can. Field-Nat.* 97(3):299–302.

Banfield, A. W. F. 1974. *The Mammals of Canada.* Buffalo, NY: University of Toronto Press.

Bannikov, A. G. 1964. "Biologie du chien viverrin en U.R.R.S." *Mammalia* 28(1):1–39.

Barash, D. P. 1974. "Neighbor recognition in two `solitary' carnivores: the raccoon (*Procyon lotor*) and the red fox (*Vulpes fulva*)." *Science* 185:794–796.

Barbu, P. 1972. "Beiträge zum Studium des Marderhundes, *Nyctereutes procyonoides ussuriensis* Matschie, 1907, aus dem Donaudelta." *Saugetierkd. Mitt.* 20:375–405.

Barnett, B. D. 1978. "Hunting behavior and food habits of the Indian wild dog (*Cuon alpinus*)." *Am. Zool.* 18:649.

Barnett, B. D., J. A. Cohen, A. J. T. Johnsingh, and M. W. Fox. 1980. "Food habits of the Indian wild dog (*Cuon alpinus*): a preliminary analysis." *J. Bombay Nat. Hist. Soc.* 77(2):313–317.

Bartmann, W. and C. Bartmann. 1986. "Mähnenwölfe (*Chrysocyon brachyurus*) in Brasilien—ein Freiland-Bericht." *Z. Kolner Zoo* 29(4):165–176.

Bartmann, W. and L. Nordhoff. 1984. "Paarbindung und Elternfamilie beim Mähnenwolf." *Z. Kolner Zoo* 27(2):63–71.

Bates, M. 1944. "Notes on a captive *Icticyon*." *J. Mammal.* 25:152–154.

Bekoff, M. 1975. "Social behavior and ecology of the African Canidae: a review." In *The Wild Canids* (M. W. Fox, ed.), pp. 120–142. New York: Van Nostrand Reinhold.

Bekoff, M. 1977. "*Canis latrans.*" *Mamm. Species* 79:1–9.

Bekoff, M. (ed.). 1978. "Coyotes: Biology, Behavior, and Management." New York: Academic Press.

Bekoff, M. and J. Diamond. 1976. "Precopulatory and copulatory behavior in coyotes." *J. Mammal.* 57(2):372–375.

Bekoff, M., J. Diamond, and J. B. Mitton. 1981. "Life-history patterns and sociality in canids: body size, reproduction and behavior." *Oecologia* 50:386–390.

Bekoff, M. and M. C. Wells. 1980. "The social ecology of coyotes." *Sci. Am.* 242:130–148.

Bekoff, M. and M. C. Wells. 1986. "Social ecology and behavior of coyotes." In *Advances in the Study of Behavior* (J. S. Rosenblatt, C. Beer, M.-C. Busnel, and P. J. B. Slater, eds.), pp. 251–338. Orlando, FL: Academic Press.

Belyaev, D. K. and L. N. Trut. 1975. "Some genetic and endocrine effects of selection for domestication in silver foxes." In *The Wild Canids* (M. W. Fox, ed.), pp. 416–426. New York: Van Nostrand Reinhold.

Benson, S. B. 1948. "Decoying coyotes and deer." *J. Mammal.* 29(4):406–409.

Bergerud, A. T. and W. B. Ballard. 1989. "Wolf predation on the Nelchina caribou herd: a reply." *J. Wildl. Manage.* 53(1):251–259.

Bergerud, A. T. and J. B. Snider. 1988. "Predation in the dynamics of moose populations: a reply." *J. Wildl. Manage.* 52:559–564.

Berry, M. P. S. 1978. *Aspects of the ecology and behaviour of the bat-eared fox,* Otocyon megalotis *(Desmarest, 1822), in the upper Limpopo River Valley.* M.S. Thesis, University of Pretoria, Pretoria.

Berry, M. P. S. 1981. "Stomach contents of bat-eared foxes, *Otocyon megalotis,* from the northern Transvaal." *S. Afr. J. Wildl. Res.* 11(1):28–30.

Berta, A. 1982. "*Cerdocyon thous.*" *Mamm. Species* 186:1–4.

Berta, A. 1984. "The Pleistocene bush dog *Speothos pacivorus* (Canidae) from the Lagoa Santa Caves, Brazil." *J. Mammal.* 65(4):549–559.

Berta, A. 1986. "*Atelocynus microtis.*" *Mamm. Species* 256:1–3.

Berta, A. 1987. "Origin, diversification, and zoogeography of the South American Canidae." In *Studies in Neotropical Mammalogy: Essays in Honor of Philip Hershkovitz* (B. D. Patterson and R. M. Timms, eds.), pp. 455–471. *Fieldiana, Zool., N.S.* 39:506 pp.

Bharucha, E., K. Asher, and R. Jugtawa. 1989. "Some interesting aspects of wolf (*Canis lupus* Linn.) behaviour observed at Guda near Jodhpur (Rajasthan)." *J. Bombay Nat. Hist. Soc.* 86(2):237–238.

Biben, M. 1982a. "Ontogeny of social behaviour related to feeding in the crab-eating fox (*Cerdocyon thous*) and the bush dog (*Speothos venaticus*)." *J. Zool., London* 196:207–216.

Biben, M. 1982b. "Object play and social treatment of prey in bush dogs and crab-eating foxes." *Behaviour* 79:201–211.

Biben, M. 1982c. "Urine-marking during agonistic encounters in the bush dog (*Speothos venaticus*)." *Zoo Biol.* 1:359–362.

Biben, M. 1983. "Comparative ontogeny of social behaviour in three South American canids, the maned wolf, crab-eating fox, and bush dog: implications for sociality." *Anim. Behav.* 31:814–826.

Bider, J. R. and P. G. Weil. 1984. "Dog, *Canis familiaris,* killed by a coyote, *Canis latrans,* on Montreal Island, Quebec." *Can. Field-Nat.* 98(4):498–499.

Birdseye, C. 1956. "Observations on a domesticated Peruvian desert fox, *Dusicyon.*" *J. Mammal.* 27(2):284–287.

Bisbal, F. J. 1988. "A taxonomic study of the crab-eating fox, *Cerdocyon thous,* in Venezuela." *Mammalia* 52(2):181–186.

Bolton, M. 1973. "Notes on the current status and distribution of some large mammals in Ethiopia (excluding Eritrea)." *Mammalia* 37(4):562–586.

Bothma, J. du P. 1966. "Food of the silver fox *Vulpes chama.*" *Zool. Afr.* 2(2):205–210.

Bothma, J. du P. 1971a. "Reports from the Mammal Research Unit: 2. Food habits of some Carnivora (Mammalia) from southern Africa." *Ann. Transvaal Mus.* 27(2):15–26.

Bothma, J. du P. 1971b. "Control and ecology of the black-backed jackal *Canis mesomelas* in the Transvaal." *Zool. Afr.* 6(2):187–193.

Bothma, J. du P., J. A. J. Nel, and A. Macdonald. 1984. "Food niche separation between four sympatric Namib Desert carnivores." *J. Zool., London* 202(3):327–340.

Bowen, W. D. 1978. *Social organization of the coyote in relation to prey size.* Ph.D. Diss., University of British Columbia, Vancouver.

Bowen, W. D. 1981. "Variation in coyote social organization: The influence of prey size." *Can. J. Zool.* 59:639–652.

Bowen, W. D. 1982. "Home range and spatial organization of coyotes in Jasper National Park, Alberta." *J. Wildl. Manage.* 46:201–216.

Bowen, W. D. and I. Cowan. 1980. "Scent-marking in coyotes." *Can. J. Zool.* 58(4):473–480.

Bowyer, R. T. 1987. "Coyote group size relative to predation on mule deer." *Mammalia* 51(4):515–526.

Boynton, K. L. 1970. "The desert fox." *Desert Mag. (Palm Desert, Calif.)* 33(1):24–27.

Bradley, L. C. and D. B. Fagre. 1988. "Coyote and bobcat response to integrated ranch management practices in south Texas." *J. Range Manage.* 41(4):322–327.

Brady, C. A. 1978. "Reproduction, growth and parental care in crab-eating foxes (*Cerdocyon thous*) at the National Zoological Park, Washington." *Int. Zoo Yearb.* 18:130–134.

Brady, C. A. 1979. "Observations on the behavior and ecology of the crab-eating fox (*Cerdocyon thous*)." In *Studies of Vertebrate Ecology in the Northern Neotropics* (J. F. Eisenberg, ed.), pp. 161–171. Washington, DC: Smithsonian Inst. Press.

Brady, C. A. 1981. "The vocal repertoires of the bush dog (*Speothos venaticus*), crab-eating fox (*Cerdocyon thous*), and maned wolf (*Chrysocyon brachyurus*)." *Anim. Behav.* 29:649–669.

Brady, C. A. 1982. *Social behavior in the crab-eating fox* (Cerdocyon thous), *the maned wolf* (Chrysocyon brachyurus), *and the bush dog* (Speothos venticus). Ph.D. Diss., Ohio University, Athens.

Brady, C. A. and M. R. Ditton. 1979. "Management and breeding of maned wolves *Chrysocyon brachyurus* at the National Zoological Park, Washington." *Int. Zoo Yearb.* 19:171–176.

Brambell, M. R. 1974. "Breeding fennec foxes *Fennecus zerda* at London Zoo." *Int. Zoo Yearb.* 14:117–118.

Brand, J. and L. Cullen. 1967. "Breeding the Cape hunting dog *Lycaon pictus* at Pretoria Zoo." *Int. Zoo Yearb.* 7:124–126.

Brown, J. B. 1973. *Behavioral correlates of rank in a litter of captive coyotes* (Canis latrans). M.S. Thesis, Purdue University, Lafayette, IN.

Brown, L. 1964. "Semien fox." *Africana* 2:45–48.

Brum-Zorilla, N. and A. Langguth. 1980. "Karyotype of South American Pampas fox *Pseudalopex gymnocercus* (Carnivora, Canidae)." *Experientia* 36(9):1043–1044.

Bueler, L. 1973. *Wild Dogs of the World.* New York: Stein and Day.

Buitron, D. 1977. *Social structure of a captive group of African wild dogs* (Lycaon pictus). M.S. Thesis, University of Minnesota, Minneapolis.

Bunker, C. D. 1940. "The kit fox." *Science* 92:35–36.

Burrows, R. 1968. *Wild Fox.* New York: Taplinger.

Burton, R. W. 1941. "The Indian wild dog." *J. Bombay Nat. Hist. Soc.* 41:691–715.

Cabrera, A. 1931. "On some South American canine genera." *J. Mammal.* 12:54–66.

Cabrera, A. 1957. "Catalogo de los mamiferos de America del Sur." *Rev. Mus. Argent. Cienc. Nat. "Bernardino Rivadavia"* 4(1):307 pp.

Cabrera, A. and J. Yepes. 1940. *Mamiferos Sud-Americanos.* Buenos Aires: Compania Argentina de Editores.

Cade, C. E. 1967. "Notes on breeding the Cape hunting dog *Lycaon pictus* at Nairobi Zoo." *Int. Zoo Yearb.* 7:122–123.

Camenzind, F. J. 1974. "Coyote talk." *Teton Mag.* 7:10–13, 40–42.

Camenzind, F. J. 1978a. *Behavioral ecology of coyotes* (Canis latrans) *on the National Elk Refuge, Jackson, Wyoming.* Ph.D. Diss., University of Wyoming, Laramie.

Camenzind, F. J. 1978b. "Behavioral ecology of coyotes on the National Elk Refuge, Jackson, Wyoming." In *Coyotes: Biology, Behavior, and Management* (M. Bekoff, ed.), pp. 267–294. New York: Academic Press.

Carbyn, L. N. 1975. "A review of methodology and relative merits of techniques used in field studies of wolves. 1973." *Proceedings of the First Working Meeting of Wolf Specialists and of the First International Conference on Conservation of the Wolf (D. H. Pimlott, ed.),* IUCN Publ. New Ser. Suppl. Pap. No. 43:134–142.

Carbyn, L. N. 1979. "Wolf howling as a technique to ecosystem interpretation in National Parks." In *The Behavior and Ecology of Wolves* (E. Klinghammer, ed.), pp. 458–470. New York: Garland Press.

Carbyn, L. N. 1982. "Coyote population fluctuations and spatial distribution in relation to wolf territories in Riding Mountain National Park, Manitoba." *Can. Field-Nat.* 96(2):176–183.

Carbyn, L. N. 1986. "Some observations on the behaviour of swift foxes in reintroduction programs within the Canadian Prairies." *Alberta Nat.* 16(2):37–41.

Carbyn, L. N. 1987. "Gray wolf and red wolf." In *Wild Furbearer Management and Conservation in North America* (M. Novak, J. A. Baker, M. E. Obbard, and B. Malloch, eds.), pp. 358–376. Toronto: Ontario Ministry of Natural Resources.

Carbyn, L. N. 1989a. "Coyote attacks on children in western North America." *Wildl. Soc. Bull.* 17(4):444–446.

Carbyn, L. N. 1989b. "Status of the swift fox in Saskatchewan." *Blue Jay* 47(1):41–52.

Carbyn, L. N. and P. C. Paquet. 1986. "Long distance movement of a coyote from Riding Mountain National Park." *J. Wildl. Manage.* 50(1):89.

Carley, C. J. 1979. "Status summary: the red wolf (*Canis rufus*)." *U. S. Fish Wildl. Serv. Endangered Species Rep.* No. 7, Albuquerque, NM.

Carr, W. H. 1945. "Gray fox adventures." *Nat. Hist.* 54:4–9.

Chesemore, D. L. 1967. *Ecology of the arctic fox in northern and western Alaska.* M.S. Thesis. University of Alaska, College.

Chesemore, D. L. 1968a. "Distribution and movements of white foxes in northern and western Alaska." *Can. J. Zool.* 46:849–854.

Chesemore, D. L. 1968b. "Notes on the food habits of Arctic foxes in northern Alaska." *Can. J. Zool.* 46:1127–1130.

Chesemore, D. L. 1969. "Den ecology of the arctic fox in northern Alaska." *Can. J. Zool.* 47:121–129.

Chesemore, D. L. 1975. "Ecology of the arctic fox (*Alopex lagopus*) in North America—A review." In *The Wild Canids* (M. W. Fox, ed.), pp. 143–163. New York: Van Nostrand Reinhold.

Childes, S. L. 1988. "The past history, present status and distribution of the hunting dog *Lycaon pictus* in Zimbabwe." *Biol. Conserv.* 44:301–316.

Clark, F. W. 1972. "Influence of jackrabbit density on coyote population change." *J. Wildl. Manage.* 36(2):343–356.

Clark, P., G. E. Ryan, and A. B. Czuppon. 1975. "Biochemical genetic markers in the family Canidae." *Aust. J. Zool.* 23:411–417.

Clark, T. W. and Stromberg, M. R. 1987. *Mammals in Wyoming.* Lawrence: University of Kansas Museum of National History.

Clutton-Brock, J. 1977. "Man-made dogs." *Science* 197:1340–1342.

Clutton-Brock, J., G. B. Corbet, and M. Hills. 1976. "A review of the Family Canidae, with a classification by numerical methods." *Bull. Br. Mus. (Nat. Hist.), Zool.* 29(3):117–199.

Cohen, J. A. 1977. "A review of the biology of the dhole or Asiatic wild dog (*Cuon alpinus* Pallas)." *Anim. Regul. Stud.* 1:141–158.

Cohen, J. A. 1978. *"Cuon alpinus."* *Mamm. Species* 100:1–3.

Cohen, J. A. and M. W. Fox. 1976. "Vocalizations in wild canids and possible effects of domestication." *Behav. Proc.* 1:77–92.

Cohen, J. A., M. W. Fox, A. J. T. Johnsingh, and B. D. Barnett. 1978. "Food habits of the dhole in South India." *J. Wildl. Manage.* 42(4):933–936.

Coimbra Filho, A. F. 1966. "Notes on the reproduction and diet of Azara's fox *Cerdocyon thous azarae* and the hoary fox *Dusicyon vetulus* at Rio de Janeiro Zoo." *Int. Zoo Yearb.* 6:168–169.

Collier, C. and S. Emerson. 1973. "Hand-raising bush dogs *Speothos venaticus* at the Los Angeles Zoo." *Int. Zoo Yearb.* 13:139–140.

Connell, W. 1944. "Wild dogs attacking a tiger." *J. Bombay Nat. Hist. Soc.* 44:468–470.

Creel, G. C. and W. A. Thornton. 1971. "A note on the distribution and specific status of the fox genus *Vulpes* in west Texas." *Southwest. Nat.* 15(3):402–404.

Crespo, J. A. 1975. "Ecology of the pampas gray fox and the large fox (culpeo)." In *The Wild Canids* (M. W. Fox, ed.), pp. 179–191. New York: Van Nostrand Reinhold.

Crespo, J. A. and J. M. de Carlo. 1963. 1. "Estudio Ecológico de una población de zorros colorados, *Dusicyon culpaeus culpaeus* (Molina) en el oeste de la provincia de Neuquén." *Rev. Mus. Argent. Cienc. Nat. "Bernardino Rivadavia"* 1:1–55.

Crisler, L. 1958. *Arctic Wild.* New York: Harper.

Cutter, W. L. 1958a. "Denning of the swift fox in northern Texas." *J. Mammal.* 39(1):70–73.

Cutter, W. L. 1958b. "Food habits of the swift fox in northern Texas." *J. Mammal.* 39(4):527–532.

Darwin, C. 1962. *The Voyage of the Beagle.* New York: Doubleday.

da Silveira, E. K. P. 1968. "Notes on the care and breeding of the maned wolf *Chrysocyon brachyurus* at Brasilia Zoo." *Int. Zoo Yearb.* 8:21–23.

Dathe, H. 1966. "Breeding the corsac fox *Vulpes corsac* at East Berlin Zoo." *Int. Zoo Yearb.* 6:166.

Dathe, H. and A. Pedersen. 1975. *"The corsac fox."* In Grzimek's Animal Life Encyclopedia (B. Grzimek, ed.), Vol. 12, p. 245. New York: Van Nostrand Reinhold.

Davidar, E. R. C. 1965. "Wild dogs (*Cuon alpinus*) and village dogs." *J. Bombay Nat. Hist. Soc.* 62:146–148.

Davidar, E. R. C. 1973. "Dhole or Indian wild dog (*Cuon alpinus*) mating." *J. Bombay Nat. Hist. Soc.* 70(2):373–374.

Davidar, E. R. C. 1975. "Ecology and behavior of the dhole or Indian wild dog *Cuon alpinus* (Pallas)." In *The Wild Canids* (M. W. Fox, ed.), pp. 109–119. New York: Van Nostrand Reinhold.

Davis, G. K. 1980. "Interaction between bat-eared fox and silver-backed jackal." *East Afr. Nat. Hist. Soc. (EANHS) Bull.* Sept./Oct. 1980:79.

Dayan, T., E. Tchernov, Y. Yom-Tov, and D. Simberloff. 1989. "Ecological character displacement in Saharo-Arabian Vulpes: outfoxing Bergmann's rule." *Oikos* 55(2):263–272.

Defler, T. R. 1986. "A bush dog (*Speothos venaticus*) pack in the eastern llanos of Colombia." *J. Mammal.* 67(2):421–422.

Dekker, D. 1968. "Breeding the Cape hunting dog *Lycaon pictus* at Amsterdam Zoo." *Int. Zoo Yearb.* 8:27–30.

Dekker, D. 1983. "Denning and foraging habits of red foxes, *Vulpes vulpes*, and their interaction with coyotes, *Canis latrans*, in central Alberta, 1972–1981." "Can. Field-Nat. 97(3):303–306.

Dekker, D. 1985. "Responses of wolves, *Canis lupus*, to simulated howling on a homesite during fall and winter in Jasper National Park, Alberta." *Can. Field-Nat.* 99(1):90–93.

Dekker, D. 1989. "Population fluctuations and spatial relationships among wolves, *Canis lupus*, coyotes, *Canis latrans*, and red foxes, *Vulpes vulpes*, in Jasper National Park, Alberta." *Can. Field-Nat.* 103(2):261–264.

Dekker, D. 1990. "Population fluctuations and spatial relationships among wolves, coyotes, and red foxes in Jasper National Park." *Alberta Nat.* 20(1):15–20.

Desai, A. A., N. Sivagenesan, and S. R. Kumar. 1988. "Interaction between dholes (*Cuon alpinus*) and a python (*Python molurus*) in Mudumalai Wildlife Sanctuary, Tamil Nadu, India." *J. Bombay Nat. Hist. Soc.* 85(1):186–187.

Deutsch, L. A. 1983. "An encounter between bush dog (*Speothos venaticus*) and paca (*Agouti paca*)." *J. Mammal.* 64(3):532–533.

Dietz, J. M. 1981. *Ecology and social organization of the maned wolf* (Chrysocyon brachyurus). Ph.D. Diss., Michigan State University, East Lansing.

Dietz, J. M. 1984. "Ecology and social organization of the maned wolf (*Chrysocyon brachyurus*)." *Smithson. Contrib. Zool.* No. 392:1–5.

Dietz, J. M. 1985. "*Chrysocyon brachyurus.*" *Mamm. Species* 234:1–4.

Dobie, J. F. 1961. *The Voice of the Coyote.* Lincoln: University of Nebraska Press.

Dolnick, E. H., R. L. Medford, and R. J. Schied. 1976. *Bibliography on the Control and Management of the Coyote and Related Canids with Selected References on Animal Physiology, Behaviour, Control Methods and Reproduction.* Beltsville, MD: Agricultural Research Service.

Doncaster, C. P., C. R. Dickman, and D. W. Macdonald. 1990. "Feeding ecology of red foxes (*Vulpes vulpes*) in the city of Oxford, England." *J. Mammal.* 71(2):188–194.

Dorst, J. and Dandelot, P. 1969. *A Field Guide to the Larger Mammals of Africa*. Boston: Houghton Mifflin.

Dragoo, J. W., J. R. Choate, and T. P. O'Farrell. 1986. "Intrapopulational variation of cranial measurements in the San Joaquin kit fox, Naval Petroleum Reserve #1, Kern County, California." *EG & G Energy Measurements. National Technical Information Service.*

Dragoo, J. W., J. R. Choate, and T. P. O'Farrell. 1987. "Intrapopulational variation in two samples of arid-land foxes." *Tex. J. Sci.* 39(3):223–232.

Dragoo, J. W., J. R. Choate, T. L. Yates, and T. P. O'Farrell. 1990. "Evolutionary and taxonomic relationships among North American arid-land foxes." *J. Mammal.* 71(3):318–332.

Drewek, J. J. 1980. *Behavior, population structure, parasitism, and other aspects of coyote ecology in southern Arizona*. Ph.D. Diss., University of Arizona, Tucson.

Drüwa, P. 1976. *Beobachtungen zum verhalten des Waldhundes* (Speothos venaticus, *Lund 1842*) *in der Gefangenschaft*. Ph.D. Diss., Rheinischen Friedrich-Wilhelms Universität, Bonn.

Drüwa, P. 1977. "Beobachtungen zur Geburt und naturlichen Aufzucht von Waldhunden (*Speothos venaticus*) in der Gefangenschaft." Zool. Gart., N.F., Jena 47(2):109–137.

Drüwa, P. 1986. "Maintaining maned wolves and giant anteaters *Chrysocyon brachyurus* and *Myrmechophaga tridactyla* together in one enclosure." *Int. Zoo Yearb.* 24/25:271–274.

Dulaney, M. 1981. "Successful birth and rearing of fennec foxes at the Cincinnati Zoo." *Anim. Keeper's Forum* 81(6):147–148.

Dunbar, R. and P. Dunbar. 1974. "Mammals and birds of the Simien Mountains National Park." *Walia* 5:4–5.

Duran, J. C. and P. E. Cattan. 1985. "The grey fox *Canis griseus* (Gray) in Chilean Patagonia (southern Chile)." *Biol. Conserv.* 34(2):141–148.

Duran, J. C., P. E. Cattan, and J. L. Yáñez. 1987. "Food habits of foxes (*Canis sp.*) in the Chilean National Chinchilla Reserve." *J. Mammal.* 68(1):179–181.

Eberhardt, L. E., R. A. Garrott, and W. C. Hanson. 1983a. "Winter movements of arctic foxes, *Alopex lagopus*, in a petroleum development area." *Can. Field-Nat.* 97(1):66–70.

Eberhardt, L. E., R. A. Garrott, and W. C. Hanson. 1983b. "Den use by arctic foxes in northern Alaska." *J. Mammal.* 64(1):97–102.

Eberhardt, W. L. 1976. "The biology of arctic and red foxes on the North Slope." *U. S. A. Proceedings, Alaska Science Conference, University of Alaska, College* 27:238.

Egoscue, H. J. 1956. "Preliminary studies of the kit fox in Utah." *J. Mammal.* 37(3):351–357.

Egoscue, H. J. 1962. "Ecology and life history of the kit fox in Tooele County, Utah." *Ecology* 43(3):481–497.

Egoscue, H. J. 1964. "The kit fox in southwestern Colorado." *Southwest. Nat.* 9(1):40–49.

Egoscue, H. J. 1975. "Population dynamics of the kit fox in western Utah." *Bull. South. Calif. Acad. Sci.* 74(3):122–127.

Egoscue, H. J. 1979. "*Vulpes velox.*" *Mamm. Species* 122:1–5.

Elliott, C. L. and R. Guetig. 1990. "Summer food habits of coyotes in Idaho's River of No Return Wilderness Area." *Great Basin Nat.* 50(1):63–65.

Emmrich, D. 1985. "Der Abessinische Wolf, *Simenia simensis* (Rüppell, 1835), Beobachtungen im Bale-Gebirge." *Zool. Gart., N.F., Jena* 55(5/6):327–340.

Estes, R. D. and J. Goddard. 1967. "Prey selection and hunting behavior of the African wild dog." *J. Wildl. Manage.* 31(1):52–70.

Ewer, R. F. 1973. *The Carnivores.* Ithaca, NY: Cornell University Press.

Failor, P. L. 1969. "Calling the gray fox." *Pa. Game News* 40(6):15–19.

Fanshawe, J. H. 1989. "Serengeti's painted wolves." *Nat. Hist.* 98(3):56–67.

Fausett, L. L. 1982. *Activity and movement patterns of the Island fox,* Urocyon littoralis, *Baird 1857 (Carnivora: Canidae).* Ph.D. Diss., University of California, Los Angeles,

Faust, R. and C. Scherpner. 1967. "A note on the breeding of the maned wolf *Chrysocyon brachyurus* at Frankfurt Zoo." *Int. Zoo Yearb.* 7:119.

Fay, F. H. and R. O. Stephenson. 1989. "Annual, seasonal, and habitat-related variation in feeding habits of the arctic fox (*Alopex lagopus*) on St. Lawrence Island, Bering Sea." *Can. J. Zool.* 67(8):1986–1994.

Feddersen-Petersen, D. 1986. "Observations on social play in some species of Canidae." *Zool. Anz.* 217(1/2):130–144.

Fentress, J. C. 1967. "Observations on the behavioral development of a hand-reared male timber wolf." *Am. Zool.* 7:339–351.

Fentress, J. C., R. Field, and H. Parr. 1978. "Social dynamics and communication." In *Behavior of Captive Wild Animals* (H. Markowitz and V. J. Stevens, eds.), pp. 67–106. Chicago: Nelson-Hall.

Ferguson, J. W. H. 1978. "Social interactions of black-backed jackals *Canis mesomelas* in the Kalahari Gemsbok National Park." *Koedoe* 21:151–162.

Ferguson, J. W. H., J. S. Galpin, and M. J. de Wet. 1988. "Factors affecting the activity patterns of black-backed jackals *Canis mesomelas.*" *J. Zool., London* 214:55–69.

Ferguson, J. W. H., J. A. J. Nel, and M. J. de Wet. 1983. "Social organization and movement patterns of black-backed jackals *Canis mesomelas* in South Africa." *J. Zool., London* 199:487–502.

Ferguson, W. W. 1981. "The systematic position of *Canis aureus lupaster* (Carnivora: Canidae) and the occurrence of *Canis lupus* in North Africa, Egypt and Sinai." *Mammalia* 45(4):459–465.

Ferrell, R. E., D. C. Morizot, J. Horn, and C. J. Carley. 1980. "Biochemical markers in a species endangered by introgression: the red wolf." *Biochem. Genet.* 18(1/2):39–49.

Field, R. 1978. *Vocal behavior of wolves* (Canis lupus): *variability in structure, context, annual/diurnal patterns, and ontogeny.* Ph.D. Diss., Johns Hopkins University, Baltimore.

Field, R. 1979. "A perspective on syntactics of wolf vocalizations." In *The Behavior and Ecology of Wolves* (E. Klinghammer, ed.), pp. 182–205. New York: Garland Press.

Fiennes, R. and Fiennes, A. 1969. *The Natural History of Dogs.* New York: Bonanza Books.

Fisher, J. L. 1981. *Kit fox diet in south-central Arizona.* M.S. Thesis, University of Arizona, Tucson.

Fisher, R. A., W. Putt, and E. Hackel. 1976. "An investigation of the products of 53 gene loci in three species of wild Canidae: *Canis lupus, Canis latrans,* and *Canis familiaris.*" *Biochem. Genet.* 14(11/12):963–974.

Flower, W. H. 1880. "On the bush dog (*Icticyon venaticus, Lund*)." *Proc. Zool. Soc. London* 1880:70–76.

Floyd, B. L. and M. R. Stromberg. 1981. "New records of the swift fox (*Vulpes velox*) in Wyoming." *J. Mammal.* 62(3):650–651.

Follman, E. H. 1978. "Annual reproductive cycle of the male gray fox." *Trans. Ill. State Acad. Sci.* 71(3):304–311.

Fox, M. W. 1970. "A comparative study of the development of facial expressions in canids; wolf, coyote and foxes." *Behaviour* 36:49–73.

Fox, M. W. 1971a. *Behaviour of Wolves, Dogs and Related Canids.* London: Jonathan Cape.

Fox, M. W. 1971b. "Possible examples of high-order behavior in wolves." *J. Mammal.* 52(3):640–641.

Fox, M. W. 1972a. "Patterns and problems of socialization in hand-reared canids: an evolutionary and ecological perspective." *Z. Tierpsychol.* 31:281–288.

Fox, M. W. 1972b. "Socio-ecological implications of individual differences in wolf litters: a developmental and evolutionary perspective." *Behaviour* 41:298–313.

Fox, M. W. (ed.). 1975. *The Wild Canids.* New York: Van Nostrand Reinhold.

Fox, M. W. 1984. *The Whistling Hunters: Field Studies of the Asiatic Wild Dog* (Cuon alpinus). Albany: State University of New York Press.

Fox, M. W. and J. A. Cohen. 1977. "Canid communication." In *How Animals Communicate* (T. Sebeok, ed.), pp. 728–748. Bloomington: Indiana University Press.

Fox, M. W. and A. J. T. Johnsingh. 1975. "Hunting and feeding in wild dogs." *J. Bombay Nat. Hist. Soc.* 72(2):321–326.

Frame, G., and L. Frame. 1981. *Swift and Enduring.* New York: E. P. Dutton.

Frame, L. H. and G. W. Frame. 1977. "Female African wild dogs emigrate." *Nature (London)* 263:227–229.

Frame, L. H., J. R. Malcolm, G. W. Frame, and H. van Lawick. 1979. "Social organization of African wild dogs (*Lycaon pictus*), on the Serengeti Plains, Tanzania, 1967–1978." *Z. Tierpsychol.* 50:225–249.

Freeman, R. C. and J. H. Shaw. 1979. "Hybridization in *Canis* (Canidae) in Oklahoma." *Southwest. Nat.* 24(3):485–499.

Fritts, S. H., W. J. Paul, and L. D. Mech. 1984. "Movements of translocated wolves in Minnesota." *J. Wildl. Manage.* 48(3):709–721.

Fritzell, E. K. 1987. "Gray fox and island gray fox." In *Wild Furbearer Management and Conservation in North America* (M. Novak, J. A. Baker, M. E. Obbard, and B. Malloch, eds.), pp. 408–420. Toronto: Ontario Ministry of Natural Resources.

Fritzell, E. K. and K. J. Haroldson. 1982. "*Urocyon cinereoargenteus.*" *Mamm. Species* 189:1–8.

Fuentes, E. R. and F. M. Jaksić. 1979. "Latitudinal size variation of Chilean foxes: tests of alternative hypotheses." *Ecology* 60(1):43–47.

Fuller, T. K. 1978. "Variable home range sizes of female gray foxes." *J. Mammal.* 59(2):446–449.

Fuller, T. K. 1989a. "Population dynamics of wolves in northcentral Minnesota." *Wildl. Monogr.* 105:1–41.

Fuller, T. K. 1989b. "Denning behavior of wolves in northcentral Minnesota." *Am. Midl. Nat.* 121:184–188.

Fuller, T. K., A. R. Biknevicius, P. W. Kat, B. van Valkenburgh, and R. K. Wayne. 1989. "The ecology of three sympatric jackal species in the Rift Valley of Kenya." *Afr. J. Ecol.* 27(4):313–323.

Fuller, T. K. and L. B. Keith. 1981. "Non-overlapping ranges of coyotes and wolves in northeastern Alberta." *J. Mammal.* 62(2):403–405.

Fuller, T. K. and B. A. Sampson. 1988. "Evaluation of a simulated howling survey for wolves." *J. Wildl. Manage.* 52(1):60–63.

Furley, C. W. 1986. "Observations on the red fox (*Vulpes vulpes arabica*) in the Al Ain Area, United Arab Emirates." *J. Bombay Nat. Hist. Soc.* 83(1):194–197.

Gangloff, L. 1972. "Breeding fennec foxes *Fennecus zerda* at Strasbourg Zoo." *Int. Zoo Yearb.* 12:115–116.

Garcia, A. 1983. "On the social behaviour of maned wolves (*Chrysocyon brachyurus*)." *Bol. Zool. (Univ. Sao Paulo)* 6:63–77.

Garrott, R. A. 1980. *Den characteristics, productivity, food habits, and behavior of arctic foxes in northern Alaska.* M.S. Thesis, Pennsylvania State University, State College.

Garrott, R. A. and L. E. Eberhardt. 1982. "Mortality of arctic fox pups in northern Alaska." *J. Mammal.* 63(1):173–174.

Garrott, R. A. and L. E. Eberhardt. 1987. "Arctic fox." In *Wild furbearer Management and Conservation in North America* (M. Novak, J. A. Baker, M. E. Obbard, and B. Malloch, eds.), pp. 394–406. Toronto: Ontario Ministry of Natural Resources.

Garrott, R. A., L. E. Eberhardt, and W. C. Hanson. 1983. "Summer food habits of juvenile arctic foxes in northern Alaska." *J. Wildl. Manage.* 47(2):540–545.

Gauthier-Pilters, H. 1962. "Beobachtungen an Feneks (*Fennecus zerda* Zimm.)." *Z. Tierpsychol.* 19(4):441–464.

Gauthier-Pilters, H. 1966. "Einige Beobachtungen über das Spielverhalten beim Fenek (*Fennecus zerda* Zimm.)." *Z. Saugetierkd.* 31:337–350.

Gauthier-Pilters, H. 1967. "The fennec." *Afr. Wildl.* 21:117–125.

Genov, P. and S. Wassilev. 1989. "Der Schakal (*Canis aureus* L.) in Bulgarien. Ein Beitrag zu seiner Verbreitung und Biologie." *Z. Jagdwiss.* 35(3):145–150.

Gese, E. M. 1990. "Reproductive activity in an old-age coyote in southeastern Colorado." *Southwest. Nat.* 35(1):101–102.

Gese, E. M., O. J. Rongstad, and W. R. Mytton. 1988a. "Relationship between coyote group size and diet in southeastern Colorado." *J. Wildl. Manage.* 52(4):647–653.

Gese, E. M., O. J. Rongstad, and W. R. Mytton. 1988b. "Home range and habitat use of coyotes in southeastern Colorado." *J. Wildl. Manage.* 52(3):640–646.

Gese, E. M., O. J. Rongstad, and W. R. Mytton. 1989. "Population dynamics of coyotes in southeastern Colorado." *J. Wildl. Manage.* 53(1):174–181.

Gier, H. T. 1975. "Ecology and behavior of the coyote (*Canis latrans*)." In *The Wild Canids* (M. W. Fox, ed.), pp. 247–262. New York: Van Nostrand Reinhold.

Gilbert, D. A., N. Lehman, S. J. O'Brien, and R. K. Wayne. 1990. "Genetic fingerprinting reflects population differentiation in the California Channel Island fox." *Nature (London)* 344(6268):764–766.

Ginsberg, J. R. and Macdonald, D. W. 1990. *Foxes, Wolves, Jackals, and Dogs. An Action Plan for the Conservation of Canids.* Gland, Switzerland: IUCN/SSC Canid Specialist Group.

Gittleman, J. L. 1984. *The behavioural ecology of carnivores.* Ph.D. Thesis, University of Sussex, Brighton, England.

Gittleman, J. L. 1985. "Functions of communal care in mammals." In *Evolution: Essays in Honor of John Maynard Smith* (P. J. Greenwood, P. H. Harvey, and M. Slatkin, eds.)., pp. 187–205. Cambridge, England: Cambridge University Press.

Goertz, J. W., L. V. Fitzgerald, and R. M. Nowak. 1975. "The status of wild Canis in Louisiana." *Am. Midl. Nat.* 93(1):215–218.

Golani, I. 1964. "Observations on the behaviour of the jackal *Canis aureus* L. in captivity." *Isr. J. Zool.* 15:28.

Golani, I. 1973. "Non-metric analysis of behavioral interaction sequences in captive jackals (*Canis aureus* L.)." *Behaviour* 44:89–112.

Golani, I. and A. Keller. 1975. "A longitudinal field study of the behavior of a pair of golden jackals." In *The Wild Canids* (M. W. Fox, ed.), pp. 303–335. New York: Van Nostrand Reinhold.

Golani, I. and H. Mendelssohn. 1971. "Sequences of precopulatory behavior of the jackal (*Canis aureus* L.)." *Behaviour* 38(176):169–192.

Golightly, R. T., Jr. and R. D. Omhart. 1984. "Water economy of two desert canids; coyote and kit fox." *J. Mammal.* 65(1):51–58.

Goszczynski, J. 1989. "Spatial distribution of red foxes *Vulpes vulpes* in winter." *Acta Theriol.* 36(26):361–372.

Gregory, W. K. 1933. "Nature's wild dog show." *Bull. N. Y. Zool. Soc.* 36(4):82–96.

Grzimek, B. 1975. "The African wild dog." In *Grzimek's Animal Life Encyclopedia* (B. Grzimek, ed.), Vol. 12, pp. 254–264. New York: Van Nostrand Reinhold.

Hall, E. R. 1981. *The Mammals of North America.* New York: John Wiley and Sons.

Hall, E. R. and Kelson, K. R. 1959. *The Mammals of North America.* New York: Ronald Press.

Hall, R. L. and Sharp, H. S. 1978. *Wolf and Man: Evolution in Parallel.* New York: Academic Press.

Haltenorth, T. and H. H. Roth. 1968. "Short review of the biology and ecology of the red fox *Canis* (*Vulpes*) *vulpes* Linnaeus 1758." *Saugetierkd. Mitt.* 16(4):339–352.

Hamilton-Smith, C. 1839. "Mammalia Vol. 18 Dogs, Vol. I." in *The Naturalist's Library* (W. Jardine, ed.). Edinburgh: W. H. Lizars.

Hamlin, K. L. and L. L. Schweitzer. 1979. "Cooperation by coyote pairs attacking mule deer fawns." *J. Mammal.* 60(4):849–850.

Harrington, F. H. 1975. *Response parameters of elicited wolf howling.* Ph.D. Diss., State University of New York, Stony Brook.

Harrington, F. H. 1981. "Urine-marking and caching behavior in the wolf." *Behavior* 76(3/4):280–288.

Harrington, F. H. 1987. "Aggressive howling in wolves." *Anim. Behav.* 35(1):7–12.

Harrington, F. H. 1989. "Chorus howling by wolves: acoustic structure, pack size and the Beau Geste effect." *Bioacoustics* 2(2):117–136.

Harrington, F. H. and L. D. Mech. 1978a. "Howling at two Minnesota wolf pack summer homesites." *Can. J. Zool.* 56:2024–2028.

Harrington, F. H. and L. D. Mech. 1978b. "Wolf vocalization." In *Wolf and Man: Evolution in Parallel* (R. L. Hall and H. S. Sharp,eds.), pp. 109–132. New York: Academic Press.

Harrington, F. H. and L. D. Mech. 1979. "Wolf howling and its role in territory maintenance." *Behaviour* 68:207–249.

Harrington, F. H. and L. D. Mech. 1982a. "Fall and winter homesite use by wolves in northeastern Minnesota." *Can. Field-Nat.* 96(1):79–84.

Harrington, F. H. and L. D. Mech. 1982b. "An analysis of howling response parameters useful for wolf pack censusing." *J. Wildl. Manage.* 46(3):686–693.

Harrington, F. H. and L. D. Mech. 1983. "Wolf-pack spacing: howling as a territory-independent spacing mechanism in a territorial population." *Behav. Ecol. Sociobiol.* 12:161–168.

Harrington, F. H., L. D. Mech, and S. H. Fritts. 1983. "Pack size and wolf pup survival: their relationship under varying ecological conditions." *Behav. Ecol. Sociobiol.* 13:19–26.

Harrington, F. H. and Paquet, P. C. (eds.). 1982. *Wolves of the World: Perspectives of Behavior, Ecology, and Conservation.* Park Ridge, NJ: Noyes.

Harrington, F. H. and J. C. Ryon. 1987. "Multiple or extended estrus in a coyote (*Canis latrans*)." *Am. Midl. Nat.* 117(1):218–219.

Harrison, D. J., J. A. Bissonette, and J. A. Sherburne. 1989. "Spatial relationships between coyotes and red foxes in eastern Maine." *J. Wildl. Manage.* 53(1):181–185.

Harrison, D. J. and J. A. Harrison. 1984. "Foods of adult Maine coyotes and their known-aged pups." *J. Wildl. Manage.* 48:922–926.

Harrison, D. L. 1968. *The Mammals of Arabia. Vol. II. Carnivora, Artiodactyla, Hyracoidea.* London: Ernest Benn.

Harrison, D. L. and P. J. J. Bates. 1989. "Observations on two mammal species new to the Sultanate of Oman, *Vulpes cana* Blanford, 1877 (Carnivora: Canidae) and *Nycteris thebaica* Geoffroy, 1818 (Chiroptera: Nycteridae)." *Bonn. Zool. Beitr.* 40(2):73–77.

Havkin, Z. and J. C. Fentress. 1985. "The form of combative strategy in interactions among wolf pups (*Canis lupus*)." *Z. Tierpsychol.* 68:177–200.

Henry, J. D. 1986. *Red Fox: The Catlike Canine.* Washington, D.C.: Smithsonian Institution Press.

Henshaw, R. E. and R. O. Stephenson. 1974. "Homing in the gray wolf (*Canis lupus*)." *J. Mammal.* 55(1):234–237.

Hershkovitz, P. 1957. "A synopsis of the wild dogs of Colombia." *Novedades Colomb.; Contrib. Cient.* 3:157–161.

Hershkovitz, P. 1961. "On the South American small-eared zorro *Atelocynus microtis* Sclater (Canidae)." *Fieldiana Zool.* 39(44):505–523.

Hersteinsson, P. 1984. *The behavioural ecology of the Arctic fox* (Alopex lagopus) *In Iceland.* Ph.D. Diss., Oxford University, Oxford.

Hersteinsson, P., A. Angerbjorn, K. Frafjord, and A. Kaikusalo. 1989. "The arctic fox in Fennoscandia and Iceland: management problems." *Biol. Conserv.* 49:67–81.

Hersteinsson, P. and D. W. Macdonald. 1982. "Some comparisons between red and arctic foxes, *Vulpes vulpes* and *Alopex lagopus*, as revealed by radio tracking." *Symp. Zool. Soc. London* 49:259–289.

Hewson, R. 1986. "Distribution and density of fox breeding dens and the effects of management." *J. Appl. Ecol.* 23:531–538.

Hill, E. P., P. W. Sumner, and J. B. Wooding. 1987. "Human influences on range expansion of coyotes in the southeast." *Wildl. Soc. Bull.* 15:521–524.

Hillman, C. N. and J. C. Sharps. 1978. "Return of swift fox to northern great plains." *Proc. S. D. Acad. Sci.* 57:154–162.

Hiruki, L. M. and I. Stirling. 1989. "Population dynamics of the Arctic fox, *Alopex lagopus*, on Banks Island, Northwest Territories." *Can. Field-Nat.* 103(3):380–387.

Hiscocks, K. and M. R. Perrin. 1987. "Feeding observations and diet of black-backed jackals in an arid coastal environment." *S. Afr. J. Wildl. Res.* 17(2):55–58.

Hiscocks, K. and M. R. Perrin. 1988. "Home range and movements of black-backed jackals at Cape Cross Seal Reserve, Namibia." *S. Afr. J. Wildl. Res.* 18(3):97–100.

Hofmann, R. K., C. F. Ponce del Prado, and K. C. Otte. 1975–1976. "Registro de dos nuevas especies de mamíferos para el Peru, *Odocoileus dichotomus* (Illiger, 1811) y *Chrysocyon brachyurus* (Illiger, 1811), con notas sobre su habitat." *Rev. Florestal Peru* 6:61–81.

Hoi-Leitner, M. and E. Kraus. 1989. "Der Goldschakal, *Canis aureus* (Linnaeus 1758), in Österreich (Mammalia austriaca 17)." *Bonn. Zool. Beitr.* 40(3/4):197–204.

Horejsi, B. L., G. E. Hornbeck, and R. M. Raine. 1984. "Wolves, *Canis lupus*, kill female black bear, *Ursus americanus*, in Alberta." *Can. Field-Nat.* 98(3):368–369.

Hornbeck, G. E. and B. L. Horejsi. 1986. "Grizzly bear, *Ursus arctos*, usurps wolf, *Canis lupus*, kill." *Can. Field-Nat.* 100(2):259–260.

Huey, L. M. 1937. "El Valle de la Trinidad, the coyote poisoner's proving ground." *J. Mammal.* 18:74–76.

Huey, R. B. 1969. "Winter diet of the Peruvian desert fox." *Ecology* 50(6):1089–1091.

Husson, A. M. 1978. *The Mammals of Suriname.* Leiden: E. J. Brill.

Ikeda, H. 1983. "Development of young and parental care of the raccoon dog *Nyctereutes procyonoides viverrinus* TEMMINCK, in captivity." *J. Mamm. Soc. Jpn.* 9(5):229–236.

Ikeda, H. 1986. "Old dogs, new treks." *Nat. Hist.* 95(8):38–45.

Ikeda, H., K. Eguchi, and Y. Ono. 1979. "Home range utilization of a raccoon dog, *Nyctereutes procyonoides viverrinus,* TEMMINCK, in a small islet in western Kyushu." *Jpn. J. Ecol.* 29:35–48.

Ilany, G. 1983. "Blandford's fox, *Vulpes cana* Blandford, 1877, a new species to Israel." *Isr. J. Zool.* 32:150. [Correct spelling for species is Blanford.]

Iriarte, J. A., J. E. Jimenez, L. C. Contreras, and F. M. Jaksić. 1989. "Small-mammal availability and consumption by the fox, *Dusicyon culpaeus,* in central Chilean scrublands." *J. Mammal.* 70(3):641–645.

Isley, T. E. and L. W. Gysel. 1975. "Sound-source localization by the red fox." *J. Mammal.* 56(2):397–404.

Jaksić, F. M., H. W. Greene, and J. L. Yáñez. 1981. "The guild structure of a community of predatory vertebrates in central Chile." *Oecologia* 49:21–28.

Jaksić, F. M., J. E. Jiménez, R. G. Medel, and P. A. Marquet. 1990. "Habitat and diet of Darwin's fox (*Pseudalopex fulvipes*) on the Chilean mainland." *J. Mammal.* 71(2):246–248.

Jaksić, F. M., R. P. Schlatter, and J. L. Yáñez. 1980. "Feeding ecology of central Chilean foxes *Dusicyon culpaeus* and *Dusicyon griseus.*" *J. Mammal.* 61(2):254–260.

Jaksić, F. M., J. L. Yáñez, and J. R. Rau. 1983. "Trophic relations of the southernmost populations of *Dusicyon* in Chile." *J. Mammal.* 64(4):693–697.

Jantschke, F. 1973. "On the breeding and rearing of bush dogs *Speothos venaticus* at Frankfurt Zoo." *Int. Zoo Yearb.* 13:141–143.

Jean, Y. and J.-M. Bergeron. 1984. "Productivity of coyotes (*Canis latrans*) from southern Quebec." *Can. J. Zool.* 62:2240–2243.

Johnsingh, A. J. T. 1978. "Some aspects of the ecology and behaviour of the Indian fox-*Vulpes bengalensis* (Shaw)." *J. Bombay Nat. Hist. Soc.* 75:397–405.

Johnsingh, A. J. T. 1982. "Reproductive and social behaviour of the dhole, *Cuon alpinus* (Canidae)." *J. Zool., London* 198:443–463.

Johnsingh, A. J. T. 1985. "Distribution and status of dhole *Cuon alpinus* Pallas, 1811 in South Asia." *Mammalia* 49(2):203–208.

Johnson, D. L. 1975. "New evidence on the origin of the fox, *Urocyon littoralis clementae,* and feral goats on San Clemente Island, California." *J. Mammal.* 56(4):925–928.

Jones, J. K., Jr., D. M. Armstrong, and J. R. Choate. 1985. *Guide to Mammals of the Plains States.* Lincoln: University of Nebraska Press.

Jones, J. K., Jr. and E. C. Birney. 1988. *Handbook of Mammals of the North-Central States.* Minneapolis: University of Minnesota Press.

Joslin, P. W. B. 1967. "Movements and home sites of timber wolves in Algonquin Park." *Am. Zool.* 7:279–288.

Keeler, C. 1975. "Genetics of behavior variations in color phases of the red fox." In *The Wild Canids* (M. W. Fox, ed.), pp. 399–413. New York: Van Nostrand Reinhold.

Kennedy, M. L. 1987. "Taxonomic status of wild canids in the southeastern United States." *Proceedings of the Third Eastern Wildlife Damage Control Conference, Gulf Shores, AL*, pp. 321–322.

Kennelly, J. J. 1978. *"Coyote reproduction."* In Coyotes: Biology, Behavior, and Management (M. Bekoff, ed.), pp. 73–93. New York: Academic Press.

Kennelly, J. J. and B. E. Johns. 1976. "The estrous cycle of coyotes." *J. Wildl. Manage.* 40(2):272–277.

Kennelly, J. J. and J. D. Roberts. 1969. "Fertility of coyote-dog hybrids." *J. Mammal.* 50(4):830–831.

Khajuria, H. 1963. "The wild dog [*Cuon alpinus* (PALLAS)] and the tiger [*Panthera tigris* (LINN.)]." *J. Bombay Nat. Hist. Soc.* 60:448–449.

Kilgore, D. L., Jr. 1969. "An ecological study of the swift fox (*Vulpes velox*) in the Oklahoma panhandle." *Am. Midl. Nat.* 81(2):512–534.

Kingdon, J. 1977. *East African Mammals: An Atlas of Evolution in Africa.* Vol. III, Part A (*Carnivores*). New York: Academic Press.

Kitchener, S. L. 1971. "Observations on the breeding of the bush dog *Speothos venaticus* at Lincoln Park Zoo, Chicago." *Int. Zoo Yearb.* 11:99–101.

Kleiman, D. G. 1967. "Some aspects of social behavior in the Canidae." *Am. Zool.* 7:365–372.

Kleiman, D. G. 1968. "Reproduction in the Canidae." *Int. Zoo Yearb.* 8:1–6.

Kleiman, D. G. 1972. "Social behavior of the maned wolf (*Chrysocyon brachyurus*) and bush dog (*Speothos venaticus*): a study in contrast." *J. Mammal.* 53(4):791–806.

Kleiman, D. G. 1977. "Monogamy in mammals." *Q. Rev. Biol.* 52:39–69.

Kleiman, D. G. and C. A. Brady. 1978. "Coyote behavior in the context of recent canid research: problems and perspectives." In *Coyotes: Biology, Behavior, and Management* (M. Bekoff, ed.), pp. 163–188. New York: Academic Press.

Kleiman, D. G. and J. F. Eisenberg. 1973. "Comparisons of canid and felid social systems from an evolutionary perspective." *Anim. Behav.* 21:637–659.

Klinghammer, E. 1979. *The Behavior and Ecology of Wolves.* New York: Garland STPM Press.

Klinghammer, E. and L. Laidlaw. 1979. "Analysis of 23 months of daily howl records in a captive grey wolf pack (*Canis lupus*)." In *The Behavior and Ecology of Wolves* (E. Klinghammer, ed.), pp. 153–181. New York: Garland STPM Press.

Koenig, L. 1970. "Zur Fortpflanzung und Jugendentwicklung des Wüstenfuchses (*Fennecus zerda* Zimm. 1780)." *Z. Tierpsychol.* 27:205–246.

Koop, K. and B. Velimirov. 1982. "Field observations on activity and feeding of bat-eared foxes (*Otocyon megalotis*) at Nxai Pan, Botswana." *Afr. J. Ecol.* 20:23–27.

Kowalski, K. 1988. "The food of the sand fox *Vulpes rueppelli* Schinz, 1825 in the Egyptian Sahara." *Folia Biol. (Krakow)* 36(1/2):89–94.

Krefting, L. W. 1969. "The rise and fall of the coyote on Isle Royale." *Naturalist* 20(4):24–31.

Krishnan, M. 1972. "An ecological survey of the larger mammals of peninsular India." *J. Bombay Nat. Hist. Soc.* 69(1):26–54.

Kruuk, H. 1972a. *The Spotted Hyena.* Chicago: University of Chicago Press.

Kruuk, H. 1972b. "Surplus killing by carnivores." *J. Zool., London* 166:233–244.

Kruuk, H. and M. Turner. 1967. "Comparative notes on predation by lion, leopard, cheetah and wild dog in the Serengeti area, East Africa." *Mammalia* 31:1–27.

Kucherenko, S. P. and V. G. Yudin. 1973. "Distribution, population density and economical importance of the raccoon dog (*Nyctereutes procyonoides*) in the Amur-Ussury district." *Zool. Zhur.* 52:1039–1045. [Russian and English Summary.]

Kühme, W. 1965a. "Communal food distribution and division of labour in African hunting dogs." *Nature (London)* 205(4970):443–444.

Kühme, W. 1965b. "Freilandstudien zur Soziologie des Hyanenhundes (*Lycaon pictus lupinus* Thomas 1902)." *Z. Tierpsychol.* 22:495–541.

Lamprecht, J. 1978. "On diet, foraging behaviour and interspecific food competition of jackals in the Serengeti National Park, East Africa." *Z. Saugetierkd.* 43:210–223.

Lamprecht, J. 1979. "Field observations on the behaviour and social system of the bat-eared fox *Otocyon megalotis* Desmarest." *Z. Tierpsychol.* 49:260–284.

Langguth, A. 1969. "Die südamerikanischen Canidae unter besonderer Berücksichtigung des Mähnenwolfes *Chrysocyon brachyurus* Illiger." *Z. Wiss. Zool.,* 179(1-2):3–188.

Langguth, A. 1975a. "Gray foxes." In *Grzimek's Animal Life Encyclopedia* (B. Grzimek, ed.), Vol. 12, pp. 267–268. New York: Van Nostrand Reinhold.

Langguth, A. 1975b. "Ecology and evolution in the South American canids." In *The Wild Canids* (M. W. Fox, ed.), pp. 192–206. New York: Van Nostrand Reinhold.

Langguth, A. 1975c. "South American canids." In *Grzimek's Animal Life Encyclopedia* (B. Grzimek, ed.), Vol. 12, pp. 268–276. New York: Van Nostrand Reinhold.

Langguth, A. 1975d. "The maned wolf." In *Grzimek's Animal Life Encyclopedia* (B. Grzimek, ed.), Vol. 12, pp. 276–279. New York: Van Nostrand Reinhold.

Laughrin, L. 1977. *The Island Fox: A Field Study of its Behavior and Ecology.* Ph.D. Diss., University of California, Santa Barbara,

Laundré, J. W. 1981. "Temporal variation in coyote vocalization rates." *J. Wildl. Manage.* 45(3):767–769.

Laundré, J. W. and B. L. Keller. 1984. "Home-range size of coyotes: a critical review." *J. Wildl. Manage.* 48(1):127–139.

Laurion, T. R. 1988. "Underdog." *Nat. Hist.* 97(9):66–71.

Lawrence, B. and W. H. Bossert. 1967. "Multiple character analysis of *Canis lupus, latrans,* and *familiaris,* with a discussion of the relationships of *Canis niger.*" *Am. Zool.* 7:223–232.

Lawrence, B. and W. H. Bossert. 1975. "Relationships of North American *Canis* shown by a multiple character analysis of selected populations." In *The Wild Canids* (M. W. Fox, ed.), pp. 73–86. New York: Van Nostrand Reinhold.

Lehner, P. N. 1978a. "Coyote communication." In *Coyotes: Biology, Behavior, and Management* (M. Bekoff, ed.), pp. 127–162. New York: Academic Press.

Lehner, P. N. 1978b. "Coyote vocalizations: a lexicon and comparisons with other canids." *Anim. Behav.* 26:712–722.

Lehner, P. N. 1982. "Differential vocal responses of coyotes to "group howl" and "group yip-howl" playbacks." *J. Mammal.* 63(4):675–679.

Leydet, F. 1977. *The Coyote: Defiant Songdog of the West.* Norman: University of Oklahoma Press.

Lindsay, I. M. and D. W. Macdonald. 1986. "Behaviour and ecology of the Ruppell's fox, *Vulpes ruppelli,* in Oman." *Mammalia* 50(4):461–474.

Lindstrom, E. 1989. "Food limitation and social regulation in a red fox population." *Holarct. Ecol.* 12:70–79.

Litvaitis, J. A. and D. J. Harrison. 1989. "Bobcat–coyote niche relationships during a period of coyote population increase." *Can. J. Zool.* 67:1180–1188.

Lloyd, H. G. 1975. "The red fox in Britain." In *The Wild Canids* (M. W. Fox, ed.), pp. 207–215. New York: Van Nostrand Reinhold.

Lloyd, H. G. 1980. "Habitat requirements of the red fox." In *The Red Fox: Symposium on Behaviour and Ecology* (E. Zimen, ed.), pp. 7–25. Boston: Biogeographica 18, Dr. W. Junk.

Lord, R. D. 1961. "A population study of the gray fox." *Am. Midl. Nat.* 66(1):87–109.

Loy, R. and J. P. Fitzgerald. 1980. "Status of the swift fox (*Vulpes velox*) on the Pawnee National Grasslands, Colorado." *J. Colo.–Wyo. Acad. Sci.* 12(1):43.

Lydeard, C. and M. L. Kennedy. 1988. "Morphologic assessment of recently founded populations of the coyote, *Canis latrans*, in Tennessee." *J. Mammal.* 69(4):773–781.

Lynch, W. 1987. "The return of the swift fox." *Can. Geogr.* 107(4):28–33.

MacCracken, J. G. and D. W. Uresk. 1984. "Coyote foods in the Black Hills, South Dakota." *J. Wildl. Manage.* 48(4):1420–1423.

MacCracken, J. G. and R. M. Hansen. 1987. "Coyote feeding strategies in southeastern Idaho: Optimal foraging by an opportunistic predator?" *J. Wildl. Manage.* 51(2):278–285.

Macdonald, D. W. 1976. "Food caching by red foxes and some other carnivores." *Z. Tierpsychol.* 42:170–185.

Macdonald, D. W. 1977. *The behavioural ecology of the red fox,* Vulpes vulpes. *A study of social organization and resource exploitation.* Ph.D. Diss., University of Oxford, Oxford.

Macdonald, D. W. 1979a. "'Helpers' in fox society." *Nature (London)* 282(5734):69–71.

Macdonald, D. W. 1979b. "The flexible social system of the golden jackal, *Canis aureus.*" *Behav. Ecol. Sociobiol.* 5:17–83.

Macdonald, D. W. 1980. "Social factors affecting reproduction amongst red foxes (*Vulpes vulpes* Linnaeus, 1758)." In *The Red Fox* (E. Zimen, ed.), Boston: Biogeographia 18, Dr. W. Junk.

Macdonald, D. W. 1981. "Resource dispersion and the social organization of the red fox (*Vulpes vulpes*)." *Worldwide Furbearer Conference Proceedings, Frostburg, Md, 1980.* (J. Chapman and D. Pursley, eds.), Vol. II, pp. 919–949.

Macdonald, D. W. (ed.). 1984. *The Encyclopedia of Mammals.* New York: Facts on File.

Macdonald, D. W., L. Boitani, and P. Barrasso. 1980. "Foxes, wolves and conservation in the Abruzzo Mountains." In *The Red Fox* (E. Zimen, ed.), pp. 223–235. The Hague: Dr. W. Junk.

Macdonald, D. W. and P. D. Moehlman. 1983. "Cooperation, altruism, and restraint in the reproduction of carnivores." In *Perspectives in Ethology* (P. Bateson and P. Klopfer, eds.), pp. 433–467. New York: Plenum Press.

Macdonald, D. W. and D. R. Voigt. 1985. "The biological basis of rabies models." In *Population Dynamics of Rabies in Wildlife* (P. J. Bacon, ed.), pp. 71–107. London: Academic Press.

MacDonald, J. T. and J. A. J. Nel. 1986. "Comparative diets of sympatric small carnivores." *S. Afr. J. Wildl. Res.* 16(4):115–121.

Mackie, A. J. and J. A. J. Nel. 1989. "Habitat selection, home range use, and group size of bat-eared foxes in the Orange Free State." *S. Afr. J. Wildl. Res.* 19(4):135–139.

MacPherson, A. H. 1969. *The Dynamics of Canadian Arctic Fox Populations.* Can. Wildl. Serv. Rep. Ser. No. 8. Ottawa: Department of Indian Affairs and Northern Development. 52 pp.

Malcolm, J. R. 1979. *Social organization and communal rearing of pups in African wild dogs* (Lycaon pictus). Ph.D. Diss., Harvard University, Boston.

Malcolm, J. R. 1980a. "Food caching by African wild dogs *(Lycaon pictus)*." *J. Mammal.* 61(4):743–744.

Malcolm, J. R. 1980b. "African wild dogs play every game by their own rules." *Smithsonian* 11(8):62–71.

Malcolm, J. R. 1986. "Socio-ecology of bat-eared foxes *(Otocyon megalotis)*." *J. Zool., London (A)* 208:457–467.

Malcolm, J. R. and K. Marten. 1982. "Natural selection and the communal rearing of pups in African wild dogs *(Lycaon pictus)*." *Behav. Ecol. Sociobiol.* 10:1–13.

Malcolm, J. R. and H. van Lawick. 1975. "Notes on wild dogs hunting zebras." *Mammalia* 39:231–240.

Manohar, B. R. and R. Mathur. 1986. "A note on the interaction of common langur *(Presbytis entellus)* and wolf *(Canis lupus)*." *J. Bombay Nat. Hist. Soc.* 83(3):653.

Marten, K. L. 1980. *Ecological sources of natural selection on animal vocalizations, with special reference to the African wild dog* Lycaon pictus. Ph.D. Diss., University of California, Berkeley.

Masopust, J. 1986. "Bat-eared fox, *Otocyon megalotis* Desmarest, 1822 and its rearing at Zoological Garden Prague." *Gazella* 13(1):105–116.

McBride, R. T. 1980. "The Mexican wolf *(Canis lupus baileyi)*: A historical review and observations on its status and distribution." *U. S. Fish Wildl. Serv. Endangered Species Rep.* No. 8, Albuquerque, NM.

McCarley, H. 1962. "The taxonomic status of wild *Canis* (Canidae) in the south central United States." *Southwest. Nat.* 7(3/4):227–235.

McCarley, H. 1975. "Long-distance vocalizations of coyotes *(Canis latrans)*." *J. Mammal.* 56(4):847–856.

McCarley, H. 1978. "Vocalizations of red wolves *(Canis rufus)*." *J. Mammal.* 59(1):27–35.

McCarley, H. 1979. "Recent changes in distribution and status of wild red wolves *(Canis rufus)*." *U. S. Fish Wildl. Serv. Endangered Species Rep.* No. 4, Albuquerque, NM.

McCarley, H. and C. Carley. 1976. "*Canis latrans* and *Canis rufus* vocalization: a continuum." *Southwest Nat.* 21(3):399–400.

McGrew, J. C. 1977. *Distribution and habitat characteristics of the kit fox (*Vulpes macrotis) *in Utah.* M.S. Thesis, Utah State University, Logan.

McGrew, J. C. 1979. "*Vulpes macrotis.*" *Mamm. Species* 123:1–6.

McNaught, D. A. 1987. "Wolves in Yellowstone?—Park visitors respond." *Wildl. Soc. Bull.* 15:518–521.

McShane, T. O. 1984. "Food of the golden jackal (*Canis aureus*) in central Niger." *Afr. J. Ecol.* 22(1):49–53.

McShane, T. O. and J. F. Grettenberger. 1984. "Food of the golden jackal (*Canis aureus*) in central Niger." *Afr. J. Ecol.* 22:49–53.

Mech, L. D. 1966. *The Wolves of Isle Royale.* U. S. Natl. Park Serv. Fauna Ser. No. 7. Washington, DC: U. S. Government Printing Office. 210 pp.

Mech, L. D. 1970. *The Wolf: The Ecology and Behavior of an Endangered Species.* Minneapolis: University of Minnesota Press.

Mech, L. D. 1974. *"Canis lupus."* *Mamm. Species* 37:1–6.

Mech, L. D. 1975. "The status of the wolf in the United States, 1973." *Wolves: Proceedings of the First Working Meeting of Wolf Specialists and of the First International Conference on Conservation of the Wolf* (D. H. Pimlott, ed.), IUCN Publ. New Ser. Suppl. Paper No. 43.

Mech, L. D. 1988. "Longevity in wild wolves." *J. Mammal.* 69(1):197–198.

Mech, L. D. and P. D. Karns. 1977. "Role of the wolf in a deer decline in the Superior National Forest." *USDA For. Serv. Res. Paper NC-148:* St. Paul, MN.

Mech, L. D. and M. E. Nelson. 1989. "Polygyny in a wild wolf pack." *J. Mammal.* 70(3):675–676.

Mech, L. D. and U. S. Seal. 1987. "Premature reproductive activity in wild wolves." *J. Mammal.* 68(4):871–873.

Medel, R. G. and F. M. Jaksić. 1988. "Ecologiá de los cánidos sudamericanos: una revision." *Rev. Chil. Hist. Nat.* 61:67–79.

Medel, R. G., J. E. Jiménez, F. M. Jaksić, J. L. Yáñez, and J. J. Armesto. 1990. "Discovery of a continental population of the rare Darwin's fox, *Dusicyon fulvipes* (Martin, 1837) in Chile." *Biol. Conserv.* 51(1):71–77.

Meester, J. and H. W. Setzer, 1971. *The Mammals of Africa: An Identification Manual.* Washington, DC: Smithsonian Inst. Press.

Mendelssohn, H., Y. Yom-Tov, G. Ilany, and D. Meninger. 1987. "On the occurrence of Blanford's fox, *Vulpes cana* Blanford, 1877, in Israel and Sinai." *Mammalia* 51(3):459–462.

Meritt, D. A., Jr. 1973. "Some observations on the maned wolf, *Chrysocyon brachyurus*, in Paraguay." *Zoologica N. Y.* 58(2):53.

Merriam, C. H. 1888. "Description of a new fox from southern California." *Proc. Biol. Soc. Washington* 4:135–138.

Meserve, P. L., E. J. Shadrick, and D. A. Kelt. 1987. "Diets and selectivity of two Chilean predators in the northern semi-arid zone." *Rev. Chil. Hist. Nat.* 60(1):93–99.

Messier, F. and C. Barrette. 1982. "The social system of the coyote (*Canis latrans*) in a forested habitat." *Can. J. Zool.* 60:1743–1753.

Mikkola, H. 1974. "The raccoon dog spreads to western Europe." *Wildlife* 16(8):344–345.

Mills, M. G. L. 1988. "Wild dogs and wild dog research in the Kruger National Park." *Quagga* 20/21:11–13.

Mitchell, R. M. 1977. *Accounts of Nepalese mammals and analysis of the host–ectoparasite data by computer techniques.* Ph.D. Diss., Iowa State University, Ames.

Mivart, S. G. 1890. *A Monograph of the Canidae.* London: R. H. Porton.

Moehlman, P. 1979. "Jackal helpers and pup survival." *Nature (London)* 277:382–383.

Moehlman, P. 1983. "Socioecology of silverbacked and golden jackals (*Canis mesomelas* and *Canis aureus*)." In *Recent Advances in the Study of Mammalian Behavior* (J. F. Eisenberg and D. G. Kleiman, eds.), Spec. Publ. No 7, pp. 423–453. Pittsburgh, Pennsylvania: American Society of Mammalogists.

Moehlman, P. 1986. "Ecology of cooperation in canids." In *Ecological Aspects of Social Evolution: Birds and Mammals* (D. I. Rubenstein and R. W. Wrangham, eds.), pp. 64–86. Princeton, NJ: Princeton University Press.

Moehlman, P. 1990. "Intraspecific variation in canid social systems." In *Carnivore Behavior, Ecology, and Evolution* (J. L. Gittleman, ed.) pp. 143–163. Ithaca, NY: Cornell University Press.

Monge-Najera, J. and B. M. Brenes. 1987. "Why is the coyote (*Canis latrans*) expanding its range? A critique of the deforestation hypothesis." *Rev. Biol. Trop.* 35(1):169–171.

Montgomerie, R. D. 1981. "Why do jackals help their parents?" *Nature (London)* 289:824–825.

Montgomery, G. G. and Y. D. Lubin. 1978. "Social structure and food habits of crab-eating fox (*Cerdocyon thous*) in Venezualan llanos." *Acta Cient. Venez.* 29:382–383.

Moore, G. C. and J. S. Millar. 1984. "A comparative study of colonizing and longer established eastern coyote populations." *J. Wildl. Manage.* 48(3):691–699.

Moore, R. E. and N. S. Martin. 1980. "A recent record of the swift fox (*Vulpes velox*) in Montana." *J. Mammal.* 61(1):161.

Morrell, S. 1972. "Life history of the San Joaquin kit fox." *Calif. Fish Game* 58(3):162–174.

Morris, P. A. and J. R. Malcolm. 1977. "The simien fox in the Balé Mountains." *Oryx* 14:151–160.

Muchmore, D. 1975. "The little swift fox." *Wyo. Wildl.* 39(7):14–15, 34.

Müller-Using, D. 1975a. "Red foxes." In *Grzimek's Animal Life Encyclopedia* (B. Grzimek, ed.), Vol. 12, pp. 245–251. New York: Van Nostrand Reinhold.

Müller-Using, D. 1975b. "African bat-eared foxes." In *Grzimek's Animal Life Encyclopedia* (B. Grzimek, ed.), Vol. 12, pp. 252–253. New York: Van Nostrand Reinhold.

Müller-Using, D. 1975c. "The fennec." In *Grzimek's Animal Life Encyclopedia* (B. Grzimek, ed.), Vol. 12 pp. 253–254. New York: Van Nostrand Reinhold.

Müller-Using, D. 1975d. "Jackals." In *Grzimek's Animal Life Encyclopedia* (B. Grzimek, ed.), Vol. 12, pp. 236–242. New York: Van Nostrand Reinhold.

Müller-Using, D. 1975e. "The red dog." In *Grzimek's Animal Life Encyclopedia* (B. Grzimek, ed.), Vol. 12, pp. 264–266. New York: Van Nostrand Reinhold.

Müller-Using, D. 1975f. "The raccoon dog." In *Grzimek's Animal Life Encyclopedia* (B. Grzimek, ed.), Vol. 12, pp. 266–267. New York: Van Nostrand Reinhold.

Murie, A. 1940. *Ecology of the Coyote in the Yellowstone.* U. S. Natl. Park Serv. Fauna Ser. No. 4. Washington, DC: U. S. Government Printing Office.

Murie, A. 1944. *The Wolves of Mount McKinley.* U. S. Natl. Park Serv. Fauna Ser. No. 5. Washington, DC: U. S. Government Printing Office.

Naaktgeboren, C. 1968. "Some aspects of parturition in wild and domestic Canidae." *Int. Zoo Yearb.* 8:6–11.

Nel, J. A. J. 1978. "Notes on the food and foraging behavior of the bat-eared fox, Otocyon megalotis." *Bull. Carnegie Mus. Nat. Hist.* 6:132–137.

Nel, J. A. J. 1984. "Behavioural ecology of canids in the southwestern Kalahari." *Koedoe, Suppl.* 27:229–235.

Nel, J. A. J., M. G. L. Mills, and R. J. van Aarde. 1984a. "Fluctuating group size in bat-eared foxes (*Otocyon m. megalotis*) in the south-western Kalahari." *J. Zool., London* 203(2):294–298.

Nel, J. A. J., M. G. L. Mills, and R. J. van Aarde. 1984b. "Fluctuating group size in bat-eared foxes (*Otocyon megalotis*), in the south-western Kalahari." *Notes Mammal Soc.* 48:294–298.

Newton, P. N. 1985. "A note on golden jackals (*Canis aureus*) and their relationship with langurs (*Presbytis entellus*) in Kanha Tiger Reserve." *J. Bombay Nat. Hist. Soc.* 82(3):633–635.

Nicholson, W. S., E. P. Hill, and D. Briggs. 1985. "Denning, pup-rearing, and dispersal in the gray fox in east-central Alabama." *J. Wildl. Manage.* 49(1):33–37.

Niewold, F. J. J. 1980. "Aspects of the social structure of red fox populations: a summary." In *The Red Fox* (E. Zimen, ed.), pp. 185–194. Boston: Biogeographica 18, Dr. W. Junk.

Northcott, T. 1975. "Long distance movement of an arctic fox in Newfoundland." *Can. Field Nat.* 89(4):464–465.

Novak, M., J. A. Baker, M. E. Obbard, and B. Malloch. 1987. *Wild Furbearer Management and Conservation in North America.* Toronto: Ontario Ministry of Natural Resources.

Novikov, G. A. 1962. *Carnivorous Mammals of the Fauna of the USSR.* Jerusalem: Israel Program Sci. Transl.

Nowak, E. 1984. "Verbreitungs- und Bestandsentwicklung des Marderhundes, *Nyctereutes procyonoides* (Gray, 1834) in Europa." *Z. Jagdwiss.* 30:137–154.

Nowak, R. M. 1978. "Evolution and taxonomy of coyotes and related *Canis.*" In *Coyotes: Biology, Behavior, and Management* (M. Bekoff, ed.), pp. 3–16. New York: Van Nostrand Reinhold.

Nowak, R. M. 1983. "A perspective on the taxonomy of wolves in North America." *Wolves in Canada and Alaska. Proceedings of the Wolf Symposium, Edmonton, Alberta, 1981* (L. N. Carbyn, ed.), *Can. Wildl. Serv. Rep. Ser.* No. 45: 10–19.

Nowak, R. M. and J. L. Paradiso. 1983. *Walker's Mammals of the World.* 4th Ed., Vol. II. Baltimore: Johns Hopkins University Press.

O'Farrell, T. P. 1984. "Conservation of the endangered San Joaquin kit fox *Vulpes macrotis mutica* on the Naval Petroleum Reserves, California." *Acta Zool. Fenn.* 172:207–208.

O'Farrell, T. P. 1987. "Kit fox." In *Wild Furbearer Management and Conservation in North America* (M. Novak, J. A. Baker, M. E. Obbard, and B. Malloch, eds.), pp. 422–431. Toronto: Ontario Ministry of Natural Resources.

Ognev, S. I. 1962. *Mammals of Eastern Europe and Northern Asia.* Vol. II. *Carnivora (Fissipedia).* Jerusalem: Israel Program Sci. Transl.

Okoniewski, J. 1982. "A fatal encounter between an adult coyote and three conspecifics." *J. Mammal.* 63(4):679–680.

Okoniewski, J. C. and R. E. Chambers. 1984. "Coyote vocal response to an electronic siren and human howling." *J. Wildl. Manage.* 48(1):217–222.

Okuzaki, M. 1979. "Reproduction of raccoon dogs, *Nyctereutes procyonoides viverrinus*, TEMMINCK, in captivity." *J. Kagawa Nutr. Coll.* 10:99–103. [Japanese with English summary.]

Ollenbach, D. C. 1930. "Notes on wild dogs in India and Burma." *J. Bengal Nat. Hist. Soc.* 4:83–86.

Ortega, J. C. 1988. "Activity patterns of different-aged coyote (*Canis latrans*) pups in southeastern Arizona." *J. Mammal.* 69(4):831–835.

Osgood, W. H. 1934. "The genera and subgenera of South American canids." *J. Mammal.* 15:45–50.

Osgood, W. H. 1943. "The mammals of Chile." *Field Mus. Nat. Hist. Publ. Zool. Ser.* 30(No. 542).

Owens, M. and D. Owens. 1984. *Cry of the Kalahari.* Boston: Houghton Mifflin.

Packard, J. M., U. S. Seal, L. D. Mech, and E. D. Plotka. 1985. "Causes of reproductive failure in two family groups of wolves (*Canis lupus*)." *Z. Tierpsychol.* 68:24–40.

Paradiso, J. L. 1968. "Canids recently collected in East Texas, with comments on the taxonomy of the red wolf." *Am. Midl. Nat.* 80(2):529–534.

Paradiso, J. L. and R. M. Nowak. 1971. *A report on the taxonomic status and distribution of the red wolf,* USDI Fish and Wildlife Service, Bureau of Sport Fisheries and Wildlife. Spec. Sci. Rept., Wildl. No. 145, Washington, DC. 36 pp.

Paradiso, J. L. and R. M. Nowak. 1972. "*Canis rufus.*" *Mamm. Species* 22:1–4.

Paradiso, J. L. and R. M. Nowak. 1982. "Wolves." In *Wild Mammals of North America: Biology, Management, and Economics* (J. A. Chapman and G. A. Feldhamer, eds.), pp. 460–474. Baltimore, MD: Johns Hopkins University Press.

Parker, G. R. and J. W. Maxwell. 1989. "Seasonal movements and winter ecology of the coyote, *Canis latrans,* in northern New Brunswick." *Can. Field-Nat.* 103(1):1–11.

Parker, W. 1988. "The Red Wolf." In *Audubon Wildlife Report 1988,* pp. 597–607. San Diego: Academic Press.

Parker, W. 1990. "Great Smoky Release to test future red wolf reintroduction changes." *Wolf* 8(4):7–8.

Pedersen, A. 1975. "Arctic fox." In *Grzimek's Animal Life Encyclopedia* (B. Grzimek, ed.), Vol. 12, pp. 243–244. New York: Van Nostrand Reinhold.

Peterson, E. A., W. C. Heaton, and S. D. Wrible. 1969. "Levels of auditory response in fissiped Carnivores." *J. Mammal.* 50:566–578.

Peterson, R. O. and R. E. Page. 1988. "The rise and fall of Isle Royale wolves, 1975–1986." *J. Mammal.* 69(1):89–99.

Petter, F. 1952. "*Le renard famélique.*" Terre Vie 1952:190–193.

Petter, F. 1957. "La reproduction du fennec." *Mammalia* 21:307–309.

Petter, G. 1964. "Origine du genre *Otocyon* (Canidae Africain de la sous-famille des Otocyoninae)." *Mammalia* 28:330–344.

Pfeffer, P. 1972. "Observations sur le comportement social et predateur du Lycaon (*Lycaon pictus*) en République Centrafricaine." *Mammalia* 36:1–7.

Phillips, R. L. 1971. "Notes on the behavior of red foxes in a large enclosure." *Proc. Iowa Acad. Sci.* 78:36–37.

Pienaar, U. de V. 1970. "A note on the occurrence of bat-eared fox *Otocyon megalotis megalotis* (Desmarest) in the Kruger National Park." *Koedoe* 13:23–27.

Pienaar, U. de V. 1973. "Predator–prey relationships amongst the larger mammals of the Kruger National Park." *Koedoe* 12:108–176.

Pils, C. M. and M. A. Martin. 1978. "Population dynamics, predator–prey relationships and management of the red fox in Wisconsin." *Wis. Dep. Nat. Resources Tech. Bull.* 105:56 pp.

Pimlott, D. H. 1975. "The ecology of the wolf in North America." In *The Wild Canids* (M. W. Fox, ed.), pp. 280–285. New York: Van Nostrand Reinhold.

Pithart, K., J. Hora, and J. Knakal. 1986. "Breeding the maned wolf, *Chrysocyon brachyurus* (Illiger, 1811) at Zoological Garden Prague." *Gazella* 13(1):63–103.

Poché, R. M., S. J. Evans, P. Sultana, M. E. Hague, R. Sterner, and M. A. Siddique. 1987. "Notes on the golden jackal (Canis aureus) in Bangladesh." Mammalia 51(2):259–269.

Pocock, R. I. 1936. "The Asiatic wild dog or dhole (Cuon javanicus)." Proc. Zool. Soc. London 1936:33–65.

Porton, I. 1983. "Bush dog urine-marking: its role in pair formation and maintenance." Anim. Behav. 31:1061–1069.

Porton, I., D. G. Kleiman, and M. Rodden. 1987. "Aseasonality of bush dog reproduction and the influence of social factors on the estrous cycle." J. Mammal. 68(4):867–871.

Prater, S. H. 1965. The Book of Indian Animals, 2nd (Rev.) Ed. Bombay: Bombay Natural History Society and Prince of Wales Museum of Western India.

Pringle, J. A. 1977. "The distribution of mammals in Natal. Part 2. Carnivora." Ann. Natal Mus. 23(1):93–115.

Pringle, L. P. 1960. "Notes on coyotes in southern New England." J. Mammal. 41(2):278.

Pulliainen, E. 1967. "A contribution to the study of the social behavior of the wolf." Am. Zool. 7:313–317.

Pulliainen, E. 1975. "Wolf ecology in northern Europe." In The Wild Canids (M. W. Fox, ed.), pp. 292–299. New York: Van Nostrand Reinhold.

Pyrah, D. 1984. "Social distribution and population estimates of coyotes in north-central Montana." J. Wildl. Manage. 48(3):679–690.

Rabb, G. B., J. H. Woolpy, and B. E. Ginsburg. 1967. "Social relationships in a group of captive wolves." Am. Zool. 7:305–311.

Ramanathan, S. 1982. "A sighting of a large dhole pack in Kanyakumari District, Tamilnadu." J. Bombay Nat. Hist. Soc. 79(3):665–666.

Ranjitsinh, M. K. 1985. "A possible sighting of Blandford's fox (Vulpes cana) in Kutch." J. Bombay Nat. Hist. Soc. 82(2):395–396. [Correct spelling is Blanford's fox.]

Rasmussen, J. L. and R. L. Tilson. 1984. "Food provisioning by adult maned wolves (Chrysocyon brachyurus)." Z. Tierpsychol. 65:346–352.

Rathbun, A. P., M. C. Wells, and M. Bekoff. 1980. "Cooperative predation by coyotes on badgers." J. Mammal. 61(2):375–376.

Redford, R. 1971, 1986. "The Language and Music of Wolves." Produced by the American Museum of Natural History. Distributed by Newman Communications Corp., Albuquerque, NM. Phono., 34 mins.

Reeder, W. G. 1949. "Aquatic activity of a desert kit fox." J. Mammal. 30(2):196.

Reich, A. 1977. "The wild dogs at Kruger Park." Afr. Wildl. 31(4):12–15.

Reich, A. 1981. The behavior and ecology of the African wild dog (Lycaon pictus) in the Kruger National Park. Ph.D. Diss., Yale University, New Haven, CT.

Ricciuti, E. R. 1978. "Dogs of war." *Int. Wildl.* 8(5):36–40.

Riley, G. A. and R. T. McBride. 1972. "A Survey of the red wolf (*Canis rufus*)." *U. S. Fish Wildl. Serv. Spec. Sci. Rep.—Wildl.* No. 162.

Robbins, J. 1986. "Wolves across the border." *Nat. Hist.* 95(5):6–15.

Roberts, A. 1951. *The Mammals of South Africa.* The Mammals of South Africa Book Fund. New York: Hafner Publishing Co. (Distributed by Central News Agency, South Africa.)

Roberts, T. J. 1977. *The Mammals of Pakistan.* London: Ernest Benn.

Robinson, W. B. 1952. "Some observations on coyote predation in Yellowstone National Park." *J. Mammal.* 33(4):470–476.

Rohwer, S. A. and D. L. Kilgore, Jr. 1973. "Interbreeding in the arid-land foxes *Vulpes velox* and *V. macrotis.*" *Syst. Zool.* 22:157–165.

Roper, T. J. and J. Ryon. 1977. "Mutual synchronization of diurnal activity rhythms in groups of Red wolf/coyote hybrids." *J. Zool., Lond.* 182:177–185.

Rosenberg, H. 1971. "Breeding the bat-eared fox *Otocyon megalotis* at Utica Zoo." *Int. Zoo Yearb.* 11:101–102.

Rosevear, D. R. 1974. *The Carnivores of West Africa.* Publ. No. 723. London: Trustees of the British Museum (Natural History).

Rothman, R. J. and L. D. Mech. 1979. "Scent-marking in lone wolves and newly formed pairs." *Anim. Behav.* 27(3):750–760.

Rowe-Rowe, D. T. 1976. "Food of the black-backed jackal in nature conservation and farming areas in Natal." *E. Afr. Wildl. J.* 14:345–348.

Rowe-Rowe, D. T. 1982. "Home range and movements of black-backed jackals in an African montane region." *S. Afr. J. Wildl. Res.* 12(3):79–84.

Rowe-Rowe, D. T. 1983. "Black-backed jackal diet in relation to food availability in the Natal Drakensberg." *S. Afr. J. Wildl. Res.* 13:17–23.

Rowe-Rowe, D. T. 1984. "Black-backed jackal population structure in the Natal Drakensberg." *Lammergeyer* 32:1–7.

Rudzinski, D. R., H. B. Graves, A. B. Sargeant, and G. L. Storm. 1982. "Behavioral interactions of penned red and arctic foxes." *J. Wildl. Manage.* 46(4):877–884.

Ryden, H. 1979. *God's Dog.* New York: Penguin Books.

Ryon, J. 1979. "Aspects of dominance behavior in groups of sibling coyote/red wolf hybrids." *Behav. Neural Biol.* 25:69–78.

Saint-Girons, M.-C. 1962. "Notes sur les dates de reproduction en captivité du fennec, *Fennecus zerda* (ZIMMERMANN 1780)." *Z. Saugetierkd.* 27:181–184.

Saint-Girons, M.-C. 1971. "Durée de vie du fennec en captivité." *Mammalia* 35:666–667.

Samuel, D. E. and B. B. Nelson. 1982. "Foxes." In *Wild Mammals of North America* (J. A. Chapman and G. A. Feldhamer, eds.), pp. 475–490. Baltimore: Johns Hopkins University Press.

Sankar, K. 1988. "Some observations on food habits of jackal (*Canis aureus*) in Keoladeo National Park, Bharatpur, as shown by scat analysis." *J. Bombay Nat. Hist. Soc.* 85(1):185–186.

Sargeant, A. B. and S. H. Allen. 1989. "Observed interactions between coyotes and red foxes." *J. Mammal.* 70(3):631–633.

Sargeant, A. B., S. H. Allen, and J. O. Hastings. 1987. "Spatial relations between sympatric coyotes and red foxes in North Dakota." *J. Wildl. Manage.* 51(2):285–293.

Schaller, G. B. 1967. *The Deer and the Tiger: A Study of Wildlife in India.* Chicago: University of Chicago Press.

Schaller, G. B. 1972. *Serengeti Lion.* Chicago: University of Chicago Press.

Schamel, D. and D. M. Tracy. 1986. "Encounters between arctic foxes, *Alopex lagopus*, and red foxes, *Vulpes vulpes*." *Can. Field-Nat.* 100(4):562–563.

Schassburger, R. M. 1978. *The vocal repertoire of the wolf: structure, function, and ontogeny.* Ph.D. Diss., Cornell University, Ithaca, NY.

Schenkel, R. 1947. "Ausdrucks-Studien an Wölfen." *Behaviour* 1:81–129.

Schenkel, R. 1967. "Submission: Its features and function in the wolf and dog." *Am. Zool.* 7:319–329.

Sclater, P. L. 1882. "Reports on the additions to the Society's menagerie in June, July, August, Setpember and October." *Proc. Zool. Soc. London* 1882:630–631.

Scott, J. P. 1950. "The social behavior of dogs and wolves: an illustration of sociobiological systematics." *Ann. N. Y. Acad. Sci.* 51:1009–1021.

Scott, J. P. 1967. "The evolution of social behavior in dogs and wolves." *Am. Zool.* 7:373–381.

Scott-Brown, J. M., S. Herrero, and J. Reynolds. 1987. "Swift fox." In *Wild Furbearer Management and Conservation in North America* (M. Novak, J. A. Baker, M. E. Obbard, and B. Malloch, eds.), pp. 432–441. Toronto: Ontario Ministry of Natural Resources.

Segal, A. N., T. V. Popovich, and M. A. Vain-Rib. 1976. "Some ecological–physiological features of the arctic fox *Alopex lagopus*." *Zool. Zh.* 55:741–754. [Russian with English summary.]

Seitz, A. 1959. "Beobachtungen an handaufgezogenen Goldschakalen (*Canis aureus algirensis* Wagner 1843)." *Z. Tierpsychol.* 16(6):747–771.

Setzer, H. W. 1961. "The canids (Mammalia) of Egypt." *J. Egypt. Public Health Assoc.* 36(3):113–118.

Shalter, M. D., J. C. Fentress, and G. W. Young. 1977. "Determinants of response of wolf pups to auditory signals." *Behaviour* 60:98–114.

Shaw, J. H. 1975. "Some contributions to the ecology and systematic position of the red wolf: an interim report." In *The Wild Canids* (M. W. Fox, ed.), pp. 278–279. New York: Van Nostrand Reinhold.

Sheldon, W. G. 1949. "Reproductive behavior of foxes in New York State." *J. Mammal.* 30:236–246.

Sheldon, W. G. 1953. "Returns on banded red and gray foxes in New York State." *J. Mammal.* 34:125.

Shortridge, G. C. 1934. *The Mammals of South West Africa.* London: William Heinemann.

Sidorov, G. N. and A. D. Botvinkin. 1987. "The corsac fox (*Vulpes corsac*) in southern Siberia." *Zool. Zh.* 66(6):914–927. [Russian with brief English abstract.]

Silver, H. and W. T. Silver. 1969. "Growth and behavior of the coyote-like canid of northern New England with observations on canid hybrids." *Wildl. Monogr.* 17:1–41.

Simonetti, J. A. 1986. "Human-induced dietary shift in *Dusicyon culpaeus*." *Mammalia* 50(3):406–408.

Simonetti, J. A. 1988. "The carnivorous predatory guild of central Chile: a human-induced community trait?" *Rev. Chil. Hist. Nat.* 61:23–25.

Simonetti, J. A., A. Poiani, and K. J. Raedeke. 1984. "Food habits of *Dusicyon griseus* in northern Chile." *J. Mammal.* 65(3):515–517.

Simonsen, V. 1976. "Electrophoretic studies on the blood proteins of domestic dogs and other Canidae." *Hereditas* 82:7–18.

Simpson, G. G. 1945. "The principles of classification and a classification of mammals." *Bull. Am. Mus. Nat. Hist.* 85:1–350.

Skead, D. M. 1974. "Incidence of calling in the black-backed jackal." *J. S. Afr. Wildl. Manag. Assoc.* 3(1):28–29.

Smithers, R. H. N. 1966. "'Matusana' a southern bat-eared fox." *Anim. Kingdom, N. Y. Zool. Soc.* 69:163–167.

Snow, C. 1967. "Some observations on the behavioral and morphological development of coyote pups." *Am. Zool.* 7:353–355.

Snow, C. 1973a. Habitat management series for endangered species. Report No. 6. San Joaquin kit fox *Vulpes macrotis mutica*." *U. S. Bur. Land Manage. Tech. Note* No. 6:23 pp.

Snow, C. 1973b. "San Joaquin kit fox, *Vulpes macrotis mutica., U. S. Bur. Land Manage. Tech. Note, Habitat Manage. Ser. Endangered Species Rep.* No. 6.

Sosnovskii, I. P. 1967. "Breeding the red dog or dhole *Cuon alpinus* at Moscow Zoo." *Int. Zoo Yearb.* 7:120–122.

Sowards, R. K. 1981. "Observations on breeding and rearing the fennec fox (*Fennecus zerda*) in captivity." *Anim. Keeper's Forum* 8(7):175–177.

Stains, H. J. 1975. "Distribution and taxonomy of the Canidae." In *The Wild Canids* (M. W. Fox, ed.), pp. 3–26. New York: Van Nostrand Reinhold.

Stanford, C. B. 1989. "Predation on capped langurs (*Presbytis pileata*) by cooperatively hunting jackals (*Canis aureus*)." *Am. J. Primatol.* 19(1):53–56.

Sterner, R. T. and S. A. Shumake. 1978. "Coyote damage-control research: a review and analysis." In *Coyotes: Biology, Behavior, and Management* (M. Bekoff, ed.), pp. 297–325. New York: Van Nostrand Reinhold.

Storm, G. L., R. D. Andrews, R. L. Phillips, R. A. Biship, D. B. Siniff, and J. R. Tester. 1976. "Morphology, reproduction, dispersal, and mortality of midwestern red fox populations." *Wildl. Monogr.* 49:81 pp.

Storm, G. L. and G. B. Montgomery. 1975. "Dispersal and social contact among red foxes: results from telemetry and computer simulation." In *The Wild Canids* (M. W. Fox, ed.), pp. 237–246. New York: Van Nostrand Reinhold.

Strickland, D. 1983. "Wolf howling in parks—the Algonquin experience in interpretation." *Wolves in Canada and Alaska. Proceedings of the Wolf Symposium, Edmonton, Alberta, 1981* (L. N. Carbyn, ed.), pp. 93–95. *Can. Wildl. Serv. Rep. Ser.* No. 45.

Stroganov, S. U. 1962. *Carnivorous Mammals of Siberia.* Jerusalem: Israel Program Sci. Transl.

Stromberg, M. R. and M. S. Boyce. 1986. "Systematics and conservation of the swift fox, *Vulpes velox*, in North America." *Biol. Conserv.* 35:97–110.

Stuart, C. T. and P. D. Shaughnessy. 1984. "Content of *Hyaena brunnea* and *Canis mesomelas* scats from southern coastal Namibia." *Mammalia* 48(4):611–612.

Sunquist, M. E. 1989. "Comparison of spatial and temporal activity of red foxes and gray foxes in north-central Florida." *Fla. Field Nat.* 17(1):11–18.

Tate, G. H. H. 1931. "Random observations on habits of South America Mammals." *J. Mammal.* 12:248–256.

Taylor, W. P. 1943. "The gray fox in captivity." *Tex. Game Fish* 1:12–13, 19.

Tembrock, G. 1963a. "Mischlaute beim Rotfuchs (*Vulpes vulpes* Linnaeus)." *Z. Tierpsychol.* 20:617–623.

Tembrock, G. 1963b. "Acoustic behavior of mammals." In *Acoustic Behavior of Animals* (R. G. Busnel, ed.), pp. 751–786. New York: Elsevier Publishing Co.

Tembrock, G. 1976. "Canid vocalizations." *Behav. Proc.* 1:57–75.

Theberge, J. B. 1966. *Howling as a means of communication in timber wolves (Canis lupus).* M.S. Thesis, University of Toronto, Toronto.

Theberge, J. B. 1990. "Potentials for misinterpreting impacts of wolf predation through prey: predator ratios." *Wildl. Soc. Bull.* 18(2):188–192.

Theberge, J. B. and J. B. Falls. 1967. "Howling as a means of communication in timber wolves." *Am. Zool.* 7:331–338.

Theberge, J. B. and C. H. R. Wedeles. 1989. "Prey selection and habitat partitioning in sympatric coyote and red fox populations, southwest Yukon." *Can. J. Zool.* 67:1285–1290.

Thenius, E. 1954. "Zur Abstammung der Rotwölfe (Gattung *Cuon* HODG-SON)." *Osterr. Zool. Z.* 5:377–378.

Thomas, O. 1918. "Some notes on the small sand-foxes of North Africa." *Ann. Mag. Nat. Hist., London* 1:242–245.

Thompson, I. D. and R. O. Peterson. 1988. "Does wolf predation alone limit the moose population in Pukaskwa Park?: A comment." *J. Wildl. Manage.* 52(3):556–559.

Thornback, J. and Jenkins, M. 1982. *The IUCN Mammal Red Data Book*, Part 1. Gland, Switzerland: Int. Union Conserv. Nat. Nat. Resour.

Thornton, W. A. and G. C. Creel. 1975. "The taxonomic status of kit foxes." *Tex. J. Sci.* 26(1/2):127–136.

Thornton, W. A., G. C. Creel, and R. E. Trimble. 1971. "Hybridization in the fox genus *Vulpes* in west Texas." *Southwest. Nat.* 15(4):473–484.

Todd, A. W., L. B. Keith, and C. A. Fischer. 1981. "Population ecology of coyotes during a fluctuation of snowshoe hares." *J. Wildl. Manage.* 45(3):629–640.

Trapp, G. R. and D. L. Hallberg. 1975. "Ecology of the gray fox (*Urocyon cinereoargenteus*): a review." In *The Wild Canids* (M. W. Fox, ed.), pp. 164–178. New York: Van Nostrand Reinhold.

Trivers, R. L. 1972. "Parental investment and sexual selection." In *Sexual Selection and the Descent of Man* (B. Campbell, ed.). Chicago: Aldine Press.

Turkowski, F. J. 1973. "The kit fox." *Natl. Parks Conserv. Mag.* 47:10–13.

Turner, K. 1968. "Bat-eared fox." *Africana* 3:27–29.

Tyler, S. 1975. "The simien fox." *Wildlife* 17(12):564–565.

Underwood, L. S. 1983. "Outfoxing the arctic cold." *Nat. Hist.* 92(12):38–47.

Uresk, D. W. and J. C. Sharps. 1986. "Denning habitat and diet of the swift fox in western South Dakota." *Great Basin Nat.* 46(2):249–253.

Valtonen, M. H., E. J. Rajakoski, and J. I. Makela. 1977. "Reproductive features in the female raccoon dog." *J. Reprod. Fertil.* 51:517–518.

van Ballenberghe, V. 1975. "Recent records of the swift fox (*Vulpes velox*) in South Dakota." *J. Mammal.* 56(2):525.

van Ballenberghe, V. 1989. "Wolf predation on the Nelchina caribou herd: a comment." *J. Wildl. Manage.* 53(1):243–250.

van Ballenberghe, V. and L. D. Mech. 1975. "Weights, growth, and survival of timber wolf pups in Minnesota." *J. Mammal.* 56:44–63.

van den Brink, F.-H. 1973. "Distribution and speciation of some carnivores—2." *Mam. Rev.* 3(3):85–95.

van der Merwe, N. J. 1953a. "The jackal." *Fauna Flora* 4:2–83.

van der Merwe, N. J. 1953b. "The coyote and the black-backed jackal." *Fauna Flora* 3:45–51.

van Gelder, R. G. 1978. "A review of canid classification." *Am. Mus. Novit.* 2646:1–10.

van Gelder, R. G. 1982. *Mammals of the National Parks*. Baltimore: Johns Hopkins University Press.

van Heerden, J. 1986. "Disease and mortality of captive wild dogs *Lycaon pictus*." *S. Afr. J. Wildl. Res.* 16(1):7–11.

van Heerden, J. and F. Kuhn. 1985. "Reproduction in captive hunting dogs *Lycaon pictus*." *S. Afr. J. Wildl. Res.* 15(3):80–84.

van Lawick-Goodall, J. and van Lawick-Goodall, H. 1971. *Innocent Killers*. Boston: Houghton Mifflin.

Vila, C., V. Urios, and J. Castroviejo. 1990. "Possible not male dominated interaction among Iberian wolves (*Canis lupus*)." *Mammalia* 54(2):312–314.

Vincent, R. E. 1958. "Observations of red fox behavior." *Ecology* 39(4):755–757.

Viro, P. and H. Mikkola. 1981. "Food composition of the raccoon dog *Nyctereutes procyonoides* Gray, 1834 in Finland." *Z. Saugetierkd.* 46:20–26.

Vogel, C. 1962. "Einige Gefangenschaftsbeobachtungen am weiblichen Fenek, *Fennecus zerda* (ZIMM. 1780)." *Z. Saugetierkd.* 27:193–204.

Voigt, D. 1987. "Red Fox." In *Wild Furbearer Management and Conservation in North America* (M. Novak, J. A. Baker, M. E. Obbard, and B. Malloch, eds.), pp. 378–392. Toronto: Ontario Ministry of Natural Resources.

Voigt, D. and W. E. Berg. 1987. "Coyote." In *Wild Furbearer Management and Conservation in North America* (M. Novak, J. A. Baker, M. E. Obbard, and B. Malloch. pp. 344–357. Toronto: Ontario Ministry of Natural Resources.

Volf, J. 1957. "A propos de la reproduction du fennec." *Mammalia* 21:454–455.

von Schantz, T. 1981. "Female cooperation, male competition, and dispersal in the red fox *Vulpes vulpes*." *Oikos* 37:63–68.

von Schantz, T. 1984. "'Non-breeders' in the red fox *Vulpes vulpes*: a case of resource surplus." *Oikos* 42:59–65.

Waithman, J. and A. Roest. 1977. "A taxonomic study of the kit fox, *Vulpes macrotis*." *J. Mammal.* 58(2):157–164.

Wakely, L. G. and F. F. Mallory. 1988. "Hierarchical development, agonistic behaviours, and growth rates in captive arctic fox." *Can. J. Zool.* 66(7):1672–1678.

Wandrey, R. 1975. "Contribution to the study of the social behaviour of captive golden jackals (*Canis aureus* L.)." *Z. Tierpsychol.* 39(1-5):365–402.

Ward, O. G. and D. H. Wurster-Hill. 1989. "Ecological studies of Japanese raccoon dogs, *Nyctereutes procyonoides viverrinus*." *J. Mammal.* 70(2):330–334.

Wasmer, D. A. 1984. "Movements and activity patterns of a gray fox in south-central Florida." *Fla. Sci.* 47(1):76–77.

Waver, R. H. 1961. "Peculiar actions of coyote and kit fox." *J. Mammal.* 42(1):109.

Wayne, R. K. and S. J. O'Brien. 1987. "Allozyme divergence within the Canidae." *Syst. Zool.* 36(4):339–355.

Wayne, R. K., B. van Valkenburgh, P. W. Kat, T. K. Fuller, W. E. Johnson, and S. J. O'Brien. 1989. "Genetic and morphological divergence among sympatric canids." *J. Hered.* 80(6):447–454.

Weiher, E. 1976. "Hand rearing fennec foxes *Fennecus zerda* at Melbourne Zoo." *Int. Zoo Yearb.* 16:200–202.

Wells, M. C. and M. Bekoff. 1982. "Predation by wild coyotes: Behavioral and ecological analyses." *J. Mammal.* 63(1):118–127.

Wendt, H. 1975a. "Kit foxes." In *Grzimek's Animal Life Encyclopedia* (B. Grzimek, ed.), Vol. 12, pp. 251–252. New York: Van Nostrand Reinhold.

Wendt, H. 1975b. "The Abyssinian jackal." In *Grzimek's Animal Life Encyclopedia* (B. Grzimek, ed.), Vol. 12, pp. 242–243. New York: Van Nostrand Reinhold.

Wenger, C. and A. T. Cringan. 1978. "Siren-elicited coyote vocalizations: an evaluation of a census technique." *Wildl. Soc. Bull.* 6:73–76.

Wenger, C. 1975. *Variations affecting siren-elicited coyote responses.* M.S. Thesis, Colorado State University, Fort Collins.

West, E. W. 1987. "Food habits of Aleutian Island arctic Foxes." *Murrelet* 68(2):93–98.

Wilcove, D. S. 1987. "Recall to the wild: Wolf reintroduction in Europe and North America." *Trends Ecol. Evol.* 2(6):146–147.

Wilson, E. O. 1975. *Sociobiology.* Cambridge, MA: Harvard University Press.

Windberg, L. A. and F. F. Knowlton. 1988. "Management implications of coyote spacing patterns in southern Texas." *J. Wildl. Manage.* 52(4):632–640.

Witt, H. 1980. "The diet of the red fox: questions about method." In *The Red Fox* (E. Zimen, ed.), pp. 65–69. Boston: Biogeographica 18, Dr. W. Junk.

Wlodek, K. and A. Krzywiński. 1986. "Zu Biologie und Verhalten des Marderhundes (*Nyctereutes procyonoides*) in Polen." *Z. Jagdwiss.* 32:203–215.

Wood, H. S. 1929. "Observations on the wild dog." *J. Bengal Nat. Hist. Soc.* 4:7–15.

Wood, J. E. 1958. "Age structure and productivity of a gray fox population." *J. Mammal.* 39(1):74–86.

Woollard, T. and S. Harris. 1990. "A behavioral comparison of dispersing and non-dispersing foxes (*Vulpes vulpes*) and an evaluation of some dispersal hypotheses." *J. Anim. Ecol.* 59(2):709–722.

Woolpy, J. 1968. "The social organization of wolves." *Nat. Hist.* 77:46–55.

Wozencraft, W. C. 1989. "Classification of the recent Carnivora." In *Carnivore Behavior, Ecology, and Evolution* (J. L. Gittleman, ed.), pp. 569–593. Ithaca, NY: Cornell University Press.

Wurster, D. H. and K. Benirschke. 1968. "Comparative cytogenetic studies in the Order Carnivora." *Chromosoma* 24:336–382.

Wyman, J. 1967. "The jackals of the Serengeti." *Animals* 10:79–83.

Yadav, R. N. 1968. "Notes on breeding the Indian wolf *Canis lupus pallipes* at Jaipur Zoo." *Int. Zoo Yearb.* 8:17–18.

Yalden, D. W., M. J. Largen, and D. Kock. 1980. "Catalogue of the mammals of Ethiopia. 4. Carnivora." *Monit. Zool. Ital. Suppl.* XIII(8):169–272.

Yamaguchi, Y. 1976. "A short note on the Japanese raccoon dogs (*Nyctereutes procyonoides viverrinus*) with stomach contents." *Bull. Kanagawa Prefect. Mus.* 9:73–76. [In Japanese with English Summary.]

Yamamoto, I. and T. Hidaka. 1984. "Utilization of `latrines' in the raccoon dog, *Nyctereutes procyonoides*." *Acta Zool. Fenn.* 171:241–242.

Young, S. P. and H. H. T. Jackson. 1951. *The Clever Coyote*. Lincoln: University of Nebraska Press.

Zeveloff, S. I. 1988. *Mammals of the Intermountain West*. Salt Lake City: University of Utah Press.

Zimen, E. 1975. "Social dynamics of the wolf pack." In *The Wild Canids* (M. W. Fox, ed.), pp. 336–362. New York: Van Nostrand Reinhold.

Zimen, E. 1976. "On the regulation of pack size in wolves." *Z. Tierpsychol.* 40:300–341.

Zimen, E. (ed.). 1980. *The Red Fox: Symposium on Behaviour and Ecology. Biogeographica 18*. Boston: Dr. W. Junk.

Zimen, E. 1981. *The Wolf: A Species in Danger*. New York: Delacorte Press.

Zumbaugh, D. M. and J. R. Choate. 1985. "Historical biogeography of foxes in Kansas." *Trans. Kansas Acad. Sci.* 88(1/2):1–13.

Zumbaugh, D. M., J. R. Choate, and L. B. Fox. 1985. "Winter food habits of the swift fox on the central high plains." *Prairie Nat.* 17(1):41–47.

INDEX